Illustrator:
Howard Chaney

Editor:
Evan D. Forbes, M.S. Ed.

Editor in Chief:
Sharon Coan, M.S. Ed.

Art Director:
Elayne Roberts

Associate Designer:
Denise Bauer

Cover Artist:
Jose Tapia

Product Manager:
Phil Garcia

Imaging:
Alfred Lau

Publishers:
Rachelle Cracchiolo, M.S. Ed.
Mary Dupuy Smith, M.S. Ed.

CE National, Inc.
1003 Presidential Dr.
P. O. Box 365
Winona Lake, IN 46590

(65) Science Experience
$35.85 set

P9-AOQ-561

Science Simulations

Intermediate

Authors:

Ruth M. Young, M.S. Ed. and Bruce T. Young, Commercial Pilot

Teacher Created Materials

Teacher Created Materials, Inc.
P.O. Box 1040
Huntington Beach, CA 92647
ISBN-1-57690-120-3

©1997 Teacher Created Materials, Inc.   Made in U.S.A.

Table of Contents

Introduction

This series of science simulations is designed for students in grades 3–5. The activities will awaken the interest of students in a variety of science areas. The simulations in this book are briefly described below.

The Life of Ants takes the students into the world of ants to find out their life cycle and how they divide up the labor within their community. Activities include observing ants in the wild to see how they behave and then examining them closely with magnifiers to study their anatomy. The life within an ant colony is enacted through the use of a shadow puppet show, illustrating all the various jobs and challenges which ants encounter. A commercial ant farm is set up in the classroom for students to make long-range observations of these fascinating insects. Finally, the unit assessment asks an open-ended question in which the students pretend that they have become worker or queen ants and must describe what this is like.

Being an Inventor enables students to develop a basic understanding of electric current generated by batteries. They begin by discovering how to light a small bulb and learning how the bulb works. This is followed by exploration on the part of the students as they invent flashlights. Next, they work with simple electric circuits to find the difference between parallel and series circuits. This leads them into finding conductors of electricity and then constructing circuit boxes. The performance-based assessment uses a series of activities taken from the unit for students to do independently, thus demonstrating their depth of understanding.

Being a Chemist takes students into the magical world of indicator dyes. Through the use of simple, easily made dyes such as red cabbage extract, students learn how chemists can distinguish between acids and bases. After experiencing this magic, students apply what they have learned to identify mystery liquids. This technique is used to help solve a problem with a hypothetical polluted lake. The final assessment is performance based, using a series of activities related to those done in the chemistry unit.

Flying a 737 enables students to take on the roles of ground and flight crew members of a commercial flight aboard a 737 jet. As the play unfolds, they learn how ground and flight crews communicate with each other, the jobs each performs for the comfort and safety of their passengers, as well as some exciting challenges which occur enroute. Activities before the play teach how planes fly and include building a working model of an airplane. Students also learn the unique alphabet and terminology used by pilots. The assessment is an open-ended question in which the student writes about a career in commercial aviation.

This wide array of scientific applications is designed to stimulate scientific curiosity. Hopefully, students' thinking will be challenged to higher and higher levels.

CE National, Inc.
1003 Presidential Dr.
P. O. Box 365
Winona Lake, IN 46590

Teaching Tips

Success with Science Simulations

The activities in *Science Simulations* have been designed to actively involve students in simulating science events and situations. Most lessons will follow the format shown below:

Title of Simulation

- Topic—states the lesson theme
- Objective—describes what students will be doing in the lesson
- Background—provides science information for students and teacher
- Materials—lists items needed for the activity
- Preparation—tells what needs to be done before conducting this lesson
- Procedure—explains how to conduct this activity
- Discussion—offers ideas for student discussion to enhance understanding
- Follow-up—provides extender idea(s) for the lesson

Storing Simulation Materials

Create a kit to preserve the materials used for each simulation so they can be reused with future students. Store items used for individual activities in heavy-duty resealable baggies. Label the baggie with the activity title and number of pieces in each baggie to avoid losing any of them. A loose-leaf notebook can be used to store lesson plans and transparencies. All of these can be placed in a large plastic storage box and labeled "Science Simulations." Supplemental items (i.e., magazine articles) can be added to the kit to use when these simulations are conducted with future classes.

Cooperative Learning Teams

Students should simulate the cooperation practiced by scientific teams that combine their efforts to make great discoveries. The activities are designed to encourage student cooperation so they can enhance the learning process by working together. Teams can be randomly assigned or designated by the teacher. Groups should be reassigned for different activities to provide the opportunity to share ideas with a variety of classmates. Team products should be shared with the entire class; thus, they need to be scientifically accurate and presented in an interesting manner. Resources, including books and periodicals, should be made available for students. These resources may be supplemented by encouraging students to bring appropriate materials from home.

Evaluation

Each of the simulation sections includes an alternative assessment as a culminating activity. The assessment can be done by teams of students or individually. Conceptual understanding and skill development is the emphasis throughout the simulations, not factual knowledge. Thus, the assessments require students to apply what they have learned in a way which will indicate their depth of understanding and their ability to use higher level thinking skills.

Consider having students collect their assignments from the activities in notebooks or folders to create portfolios. They should evaluate their work by comparing it with previous assignments as they add to their portfolios. Have each student write a critique of his/her work at the end of each simulation section and add it to his/her portfolio. Let each student choose one assignment from each simulation section to submit as part of his/her final grade.

The Life of Ants

Scientists think ants gradually developed from wasps more than 100 million years ago. They resemble wasps more than any other insect. In fact, the red insect called "velvet ant" is actually a wasp. Ants have bumps at the tops of their waists, which wasps don't have.

There are at least 10,000 different species of ants. Most are drab colors such as brown, black, or rust. However, some are green, yellow, blue, or purple. They range in size from 1" (2.5 cm) to .25" (1 mm). Most ants are extraordinarily strong, however; some can lift objects weighing 10 to 50 times their own weight.

Ants are considered the most highly developed social insects. Other social insects include bees, termites, and some wasps. Ants live in communities called **colonies**, which can be underground, in dirt mounds, inside trees, or in hollow parts of other plants. Army ants do not have permanent nests. Some ants raid the nests of other ants, stealing the young and raising them as their slaves. Harvester ants gather seeds and store them inside their nests. Dairy ants keep insect larvae that give off sweet liquid when the ants "milk" them.

These fascinating insects are found all over the world, except for the Arctic and Antarctic.

Life Cycle of the Ant

A queen ant may lay one egg every 10 minutes of her life. The tiny eggs are licked clean by worker ants. After a few weeks, they hatch into helpless, legless larvae which are fed by the workers. Eventually, the larvae spin silken cocoons and enter the pupa stage. The pupae rest inside their cocoons and undergo metamorphosis (change) into adults. When they are fully developed adults, workers usually tear open the cocoons to help them out. These adults are usually small, wingless female workers. Sometimes they are large-winged queens and males. The males will fertilize the queens who will return to their colonies to lay eggs or begin a new colony.

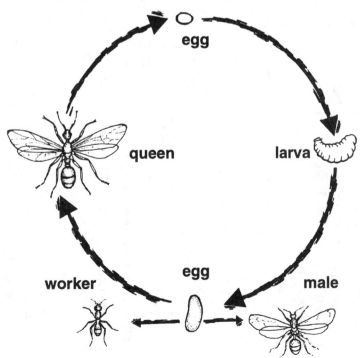

Ant Anatomy

Highly Magnified View of Head

The head has a pair of jaws called **mandibles,** which move from side to side. They are used to grasp food, carry young, and fight enemies. Two antennae on the head are used to smell, touch, taste, and hear. Two large compound eyes are made up of 6–1,000 lenses. Each lens sees a small part of an image. When all of these lenses are put together they form a whole picture, which is not very clear. The eye easily sees movement, however.

Inside an Ant's Body

The ant's **brain** inside its head is connected to a nerve cord which runs through the trunk and metasoma. Nerves branch out from the nerve cord to all parts of its body. The **heart** is a long tube stretching to the rear of the gaster. Muscle contractions force blood forward through the tube. Blood empties out of the tube through one-way openings and flows throughout the body, bathing all tissues and organs. The blood is colorless. There are no **lungs,** but oxygen enters the body through tiny holes along the sides of the body and is carried through tubes to all parts of the body. Carbon dioxide is collected and passes out through the same openings (spiracles). The **digestive system** consists of a mouth which chews food into a liquid which travels through a tube to the **crop,** a pouch. This food is stored in the crop to share with other ants; some of it passes to the **stomach** and is digested. Some ants have a sting which delivers poison from a gland.

Outside View of Body

The ant, like all insects, has three body parts; head, trunk (thorax), and metasoma (abdomen) and six legs attached at the trunk.

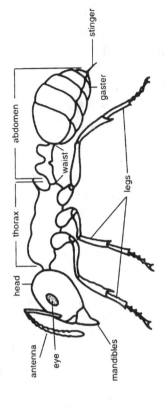

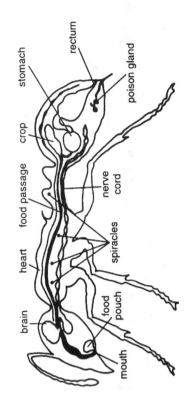

6

Observing Ants in the Wild

Topic: Ant activity

Objective: Students will learn how wild ants search for food by making observations of them.

Materials:

- pieces of food such as fruit, candy, or meat
- popcorn and a popcorn maker

Preparation:

- Find a location on the school grounds where ants can be observed, such as a sandbox or dirt.
- Spend some time observing the ants. Scatter some of the food near the ants and draw a line through their trail to watch how they behave.
- On the day of this lesson, scatter food near the ant area to attract them, if they are not already active. You may want students to bring their chairs so they can sit in small groups near the ants and make careful observations. Consider letting students videotape the ants so they can make closer observations back in the classroom.
- Make fresh popcorn and place it in the classroom where it is hidden but where students will smell it as they enter the room.

Procedure:

- As the students enter the room and smell the popcorn, ask them to see if they can find it. Once they discover it, serve the popcorn to them.
- Take the students to the ant area and divide them into small groups to observe and discuss what they see the ants doing. Tell the students to watch one ant for a few minutes to see how it moves. Ask them what happens when one ant meets another.
- Have them watch the ants' behavior when food is placed near them. Tell them to see if the food is carried off and to observe how this is done.
- Look for an uninterrupted trail of ants and make several breaks in it by drawing lines through it with at stick at 6" (15 cm) intervals. Tell students to watch how the ants rediscover the path.

For Discussion:

- Explain that ants have very poor vision so they can not see the path. After a while, remind students of how they found the popcorn in the classroom (by smell).
- Explain that as the ants travel, they leave a smell behind them which is followed by the ants behind them. They use their antennae to smell. When the line is drawn across their path, the smell is broken and they need to move around until they find it again.

Follow Up:

Have students write a brief description of what they observed the ants doing. Be sure they illustrate the essay to give more details.

Ants Close Up

Topic: Body structure and movements of an ant

Objective: Students will learn about the body structure and movements of a live ant through observations.

Materials:

- live ants inside resealable baggies (one per group)
- copies of Ant Record data sheet (page 9) for each student
- resealable baggies
- magnifier for each group
- spoon
- live ants from the wild
- overhead projector

Preparation:

- This lesson should be conducted the day after observing the ants in the wild.
- Use the spoon to place several ants in the resealable baggies. Blow a bit of air into the bag before sealing it.

Procedure:

- Discuss the observations students made of wild ants in the earlier lesson (page 7).
- Place one of the baggies containing live ants on the overhead projector. Let students observe and discuss what they see for a few minutes.
- Tell the students that they are going to work in groups to observe ants that are inside baggies.
- Divide the students into groups and distribute the ants in the baggies, magnifiers, and Ant Record sheet.
- Show a copy of the record sheet on the overhead and discuss it with the students.
- Remind them that they are working with live insects, which are much smaller than they are and therefore need to be handled very gently. Explain that following this lesson, they will go outside to release the ants where they were collected. Caution the students not to try to open the bags, for the safety of the ants. Let them observe and record their data on the record sheet.

For Discussion:

- Place one of the baggies on the overhead projector so students can observe it as a class. Have a spokesperson from each group share something they noticed about the ants which had not been mentioned at the beginning of this lesson.
- Review the transparency of the Ant Anatomy (page 6). Have students compare their drawings to see if they found most of these external parts of the ant.

Follow Up:

Perform the shadow puppet show which simulates life in an ant colony (pages 10–21).

Ants Close Up *(cont.)*

Ant Record

Name: _____

Make careful observations of one ant with a magnifier and then draw what you see.

Ant viewed from above   **Ant viewed from the side**

Describe how your ant walks.

_____

_____

Make a series of drawings to show how the ant moves when walking.

1	2	3	4

Tell what you observed of the ants in the wild. Be sure to describe how they reacted to the food.

_____

_____

Draw a picture to show one of the ants in action.

Tell what the ant is doing in your picture:_____

Life in an Ant Colony

Topic: Life in an ant colony

Objective: Students will learn about life in an ant colony by performing a shadow puppet show simulation.

Background: Share the information about the life cycle and anatomy of ants (pages 5 and 6) with the students.

Materials:

- transparencies of the puppets, props, and scenery (pages 17–21)
- transparency of the inside view of an ant (page 6)
- white butcher paper
- glue gun and restickable glue stick
- copies of the play script (page 12–16)
- overhead projector
- about 10 rice grains (to represent ant eggs)
- *14' (1.5 m) of 22- or 24-gauge wire
- **three-sided display board 36" x 36" (90 cm x 90 cm) or three large, heavy-duty cardboard pieces

  *wire may be purchased at a hardware store or a craft store

  **available at most office supply stores

Procedure:

- Assign students to the read the script; they must be all girls' voices since the queen and workers are all female. Let them practice reading the script slowly to permit the puppeteers to maneuver their puppets as described in the story.
- Select students to be the puppeteers; these may be both boys and girls.
- Choose two students to be stagehands to operate the overhead projector and put the scenery and props on it as called for in the script.
- Other students will be needed to set up the stage and to create a program for the play.
- Practice the play so cast members learn their roles and become accustomed to working with the puppets and the scenery projected onto the screen.
- Perform the play for the classroom. Have students exchange roles so all students have the opportunity to participate.

Follow Up:

- Let students add lines to the script after doing more research with ants (see lesson on page 8).
- Perform the play for other students in the school. A larger screen can be made for larger audiences. Use a bedsheet hanging from the ceiling or other support as a screen. Move the overhead projector back further to enlarge the scenery. Increase the sizes of the puppets by making larger copies on the transparencies and moving them further away from the screen.

Life in an Ant Colony *(cont.)*

Synopsis:

The life in an ant colony is depicted beginning with the queen ant constructing a new colony after her mating flight. She digs out an underground chamber, seals herself into it, and lays eggs. When the eggs hatch, she cares for the larvae until they become adult worker ants. These workers, all female, leave the nest to forage for food, while the queen continues to lay eggs. The workers now take over the responsibility of caring for the eggs and larvae, building and repairing the colony, and supplying food for its occupants. Finally, another queen and male ants develop in the mature colony and leave in their mating flight. This ultimately results in the construction of more colonies, as well as expanding the one from which they originated.

Staging:

The shadow puppet show is performed using the shadows of puppets and scenery cast on white paper. The puppet stage can be made by using a three-sided display board or large cardboard.

Preparation:

- Make a three-sided puppet stage from cardboard or use a display board. The front of the stage should be at least 36" (90 cm) square. Cut at least a 34" (85 cm) square near the top. Cover this hole with white butcher paper. Place the stage on a table and cover the table so puppeteers can be hidden behind it.

- Cut out the puppets and scenery printed on transparencies. Use a glue gun to attach the puppets to the end of 1' (30 cm) wire so they may be held vertically.

- Place the first scenery transparency on the stage of the overhead and project the image onto the paper screen so that it fills the opening. Put the shadow of one of the puppets on the screen with the background and move the projector to enlarge or decrease the size of the picture so it fits the sizes of the puppets.

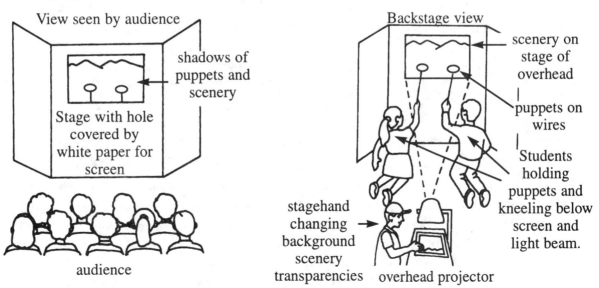

Life in an Ant Colony *(cont.)*

Puppet Show Script

(Puppets: the winged queen ant and males; Scenery: none; Action: ants flying)

Queen: Hi! I'm a queen ant. I've just left my home, which you call a colony. It feels wonderful to have these wings and fly so high. Those smaller winged ants are males. This is our mating flight.

(Puppet: queen ant, missing one wing; Scenery: ground above queen's chamber)

Queen: Now that the mating flight is over, I won't ever need these wings again so I'll rub them off. Too bad; I will miss the fun of flying, but my job now is to build a new colony.

(Puppet: queen ant without wings; Scenery: gradually uncover queen's chamber as she digs; Props: rice representing eggs)

Queen: Watch me dig with my front legs and jaws to make this hole deep enough so I will have a chamber where I can lay eggs which will become the workers in my colony. I need to seal myself inside this chamber to keep out the rain and anything which might want to eat me. I won't ever leave this colony; for the rest of my life, I will be busy laying eggs and will never go outside again. I may live from 10 to 20 years and may lay thousands or even millions of eggs. Whew! That's a lot of offspring! Now that I have sealed myself in, I better get busy with my job.

(Puppets: queen, larvae; Scenery: underground chamber; Props: more rice and larvae)

Narrator: Since the queen doesn't go out to look for food, she lives off her body fat and gets some nourishment from her wing muscles. She may also eat some of the smaller eggs she lays. After laying the eggs, she licks them and takes care of them. The eggs hatch several days later into tiny, colorless worms called larvae. They are helpless and can't move or care for themselves. The queen feeds them the liquid food she has stored in her crop. She forces it out of her mouth and into the mouth of the larvae. She also continues to lay more eggs. This is a very busy queen ant!

(Puppets: queen, 3 workers; Scenery: queen's chamber; Props: more rice, larvae, pupa)

Narrator: Within a few weeks, the larvae have eaten a lot and grown larger. They now begin to spin a cocoon of very thin silk around themselves. This stage is called the pupa. The larvae now go through a great change, or metamorphosis, and become adult ants. When they are ready to get out of the cocoon, the queen will bite open the silk to help them emerge.

Queen: Now my family is getting bigger. I have laid many more eggs, larvae are hatching, and, at last, the workers are beginning to emerge from their cocoons. They will take over the work of building more chambers in this colony, gathering food for the larvae and me, and taking care of the eggs and larvae. I can now spend all my time laying eggs.

(Puppets: 3 workers, spider; Scenery: ground above ant colony; Props: larvae, pupae)

Worker #1: We have been sent out to look for food for members of our colony. Our food can be anything such as a dead insect or animal, pieces of food people drop, or seeds. Once we locate the food, we will return to the nest to tell some of the other workers where to look for it. Oh look! There is a dead spider; it will be a lot of food for our colony. We need to rush back to let the others know where the food is and get their help to carry it back to the nest.

Life in an Ant Colony *(cont.)*

PuppetShow Script*(cont.)*

(Puppets: 2 workers; Scenery: ground above colony and section #1 of the colony; Action: worker enters nest and begins pushing another worker toward the surface)

Worker #2: Why are you pushing me?

Worker #1: We have found food and need help getting it back to the colony.

(Scenery: area above colony; Action: workers leave the entrance of the nest; they begin to feel around with their antennae)

Narrator: The worker ants use their antennae to smell the pheromone or scent which the workers left for them as a trail to follow to the food.

(Action: Workers find the spider and begin to use their jaws to tug it back to the nest. They push the spider into the nest.)

Worker #1: Teamwork, girls! Use your jaws to help get this great dinner back to the colony.

Narrator: Ants can carry from 10 to 50 times their own weight in their jaws! It would be like you carrying your bed in your mouth.

(Scenery: inside view of an ant; Action: stagehand uses a pencil to point to the body parts)

Narrator: If you had x-ray vision, you could see this inside view of an ant. Each ant tears off a piece of the spider and chews it until it becomes liquid. They swallow the liquid food, and it moves to their crops located in their metasomas or abdomens. The crop is a public stomach since this food will be shared with the rest of the colony. Some food passes through the gizzard to the ant's own stomach to be digested by it.

(Puppets: 2 workers; Scenery: ground above colony and #1 entrance area; Action: one worker passes food to another)

Worker #3: Let me give you some of this liquid food from my crop. Take it to the queen and then return for more. The other workers and I have gathered plenty of food which we will pass on to the rest of the ants in our nest.

(Puppets: queen and 3 workers; Scenery: #2 queen's chamber; Props: rice and larvae; Action: 2 workers tend the eggs; other carries food to the queen and then carries out larvae)

Worker #4: Take some of this tasty spider the workers found outside, your majesty.

Queen: Thanks; it is tasty. I was getting really hungry since I've laid nearly 50 eggs today. These workers are taking good care of the eggs and me.

Worker #4: These eggs have hatched into larvae. I'll carry them to the nursery.

(Use glue to stick larvae cluster to mouth of worker # 4 puppet.)

(Puppets: 2 workers, pupae cluster; Scenery: #3 larvae nursery; Props: larvae in nursery chamber; Action: one worker caring for larvae as another carries in the small cluster of new larvae.)

Life in an Ant Colony *(cont.)*

Puppet Show Script *(cont.)*

Worker #4: Here are more larvae for you to take care of. Feed them well; we need more workers. Our colony is growing fast, and there is more work than we can do.

Worker #5: This larva has spun a silk cocoon around its body and is now a pupa. It needs to be taken to the upper-level nursery.

Worker #4: I'll carry it up there for the other workers to take care of.

(Use restickable glue to stick pupa to mouth of worker # 4 puppet.)

(Puppets: 2 workers; Scenery: #4 pupae nursery; Props: pupa)

Worker #4: Here is another pupa for you to take care of. It will take about two to three weeks for it to gradually change into an adult.

Worker #6: Look, here is a pupa ready to emerge now. Help me tear open this silk cocoon so it can crawl out. *(Workers 4 and 6 pull at the cocoon.)*

Worker #7: Ah! Free at last! That was a long sleep inside that cocoon. What do I do now?

Worker #6: First, go get something to eat and then get right to work.

Worker #4: Come with me; I'll show you where to get the food and explain your assignment. We sure are glad to have your help. You'll find that there isn't any time to rest in this colony. We are always busy taking care of everyone here, including going to get food and protecting the nest.

(Puppets: 3 worker ants; Action: new worker ant moves near entrance and is fed by another ant)

Worker #3: Welcome to our colony. Let me give you your first meal as an adult so you will get enough strength to do your work.

Worker #7: That was delicious; now I'm ready to do my job. Where do I go?

(Puppet: 3 worker ants; Scenery: #5 cave-in area; Action: worker #3 moves out of the entrance, the other workers move deeper into the colony to help dig out a chamber)

Worker #4: Oh no! I smell danger. It seems to be coming from deeper in the colony. Let's hurry to find out what has happened.

Worker #7: What do you mean? How can you "smell" danger?

Worker #4: *(Rushing down into the tunnel below).* It's called pheromone, and it comes from our body like perfume. We use it to help us recognize our family members, to send out calls for help, or lay a trail to follow outside the colony.

Worker #2: I'm so glad you came! There's been a cave-in at the entrance to this chamber, and I need your help. There is no time to lose; workers and larvae are trapped inside the chamber. Use your jaws and front legs to help dig them out. *(All three workers dig to clear away the chamber.)*

Life in an Ant Colony *(cont.)*

Puppet Show Script *(cont.)*

(Puppets: #3 worker top view; Scenery: top view of ground near colony entrance; Action: worker following pheromone trail to food source)

Worker #3: I can easily follow the trail laid down by my sister ants. They left some of their pheromone as a trail which I can smell using my antennae.

(Worker runs into footprint.)

What happened? I've lost the scent. I'll go back to where I last smelled it.

(Action: worker wanders back and forth, searching around the footprint.)

There, at last, I smell it again! Now, if I can just find the other end of the trail. I'll smell around this spot for the scent. I know it's here somewhere.

(Action: worker searches around footprint, then discovers the trail again; Scenery: dead ants in a line within the footprint)

All right! Here is the other end of the trail. There are some of my sisters, and they are smashed flat! A giant must have stepped on them. Oh, no! I can feel the ground shaking, the giant must still be walking near here. I need to rush along this trail to get the food and get back.

Worker #1: Well! It's about time you showed up. What kept you so long?

Worker #3: The trail broke off back there. I think a giant walked on some of our sisters. Did you feel that?

Worker #1: The ground shook. The giant is coming! Get everyone out of the way!

(Ants scurry around, looking for a hiding place.)

Worker #1: I think the danger is past. Let's pick up the trail and head back to the colony. Everyone pick up a seed to carry back with you.

(Ants move toward seeds to pick them up.)

(Puppets: 4 workers; Scenery: entire colony; Action: ant from another colony arrives)

Worker #7: *(at the colony entrance, smelling the ant which is trying to get into the colony)*

Oh, no you don't! You're not going to get into my colony. You don't smell like the rest of us so you aren't my sister.

(Action: picks up enemy ant with its jaws and tosses it aside; two other ants sneak past, go into the colony larvae nursery)

(Puppets: 2 workers; Action: two enemy workers steal larvae from nursery)

Narrator: Sometimes ants will raid another colony to take their larvae to raise as workers in their own colony. They may even steal eggs from another colony and use these as food.

Worker #8: Help me carry out these larvae. We need more workers in our colony so we'll take these back to our nest. *(Attach larvae to ants' jaws with restickable glue.)*

Worker #9: They sure are heavy. I think they will become healthy workers.

(Action: ants work their way out of the colony entrance and hurry away)

(Puppets: 3 workers; Scenery: entire colony; Action: ants return with seeds)

Life in an Ant Colony *(cont.)*

Puppet Show Script *(cont.)*

Worker #1: What happened here? Who killed this ant?

Worker #7: I did. It was from a nearby colony and was trying to get inside our home. Unfortunately, two of its sisters got past me and stole some of our larvae. I tried to stop them, but they were too fast for me.

Worker #1: You did a good job by stopping this one at least. You had a rough first day on the job. We know you will be a good worker.

(Puppets: workers and queen; Scenery: entire colony; Action: workers tending ants, queen laying eggs)

Narrator: Everything is back to normal inside the colony. The queen will lay more eggs which will hatch into larvae that become pupae and then emerge as adult workers. Life in the colony is always busy and sometimes dangerous. Worker ants don't have any time to play games or take vacations. They work so hard that they only live from one to five years.

(Puppets: winged queen and male ants; Action: queen and male ants in mating flight over colony)

When this colony becomes larger, another queen and male ant will emerge and then rush around inside the colony until they find the entrance. The males will fly off together with the new queen to mate.

(Puppets: male ants, spider; Scenery: ground above colony; Action: spider eating male ants, queen ant flies off to build another colony)

The males fall back to earth where they soon die. They serve as food for other animals such as spiders. The queen may return to her colony and join her mother queen in adding to the population in the colony. She may also fly off to start her own colony.

Ants have been on Earth for millions of years and will, no doubt, continue to live here for a long time to come. There are at least 10,000 more species of ants; this story is about only one type. Read books to find out about the other species of ants and to discover why they are so important to us.

16

Life in an Ant Colony *(cont.)*

Ant Puppets and Props

To the Teacher: Make transparencies of the puppets and cut them out, following the outline around them, without including the name. Only the print of the picture should cast the shadow; therefore, trim away the outline. The number of puppets needed is shown in ( ) beside the name.

Attach the wire with a hot glue gun to hold the puppets in a vertical position. Puppeteers will hold the wire from below the puppets so their bodies do not cast shadows on the screen. Puppets will be in focus when held near the screen. The ant larvae will be attached to the ant puppet with restickable glue, when called for in the script.

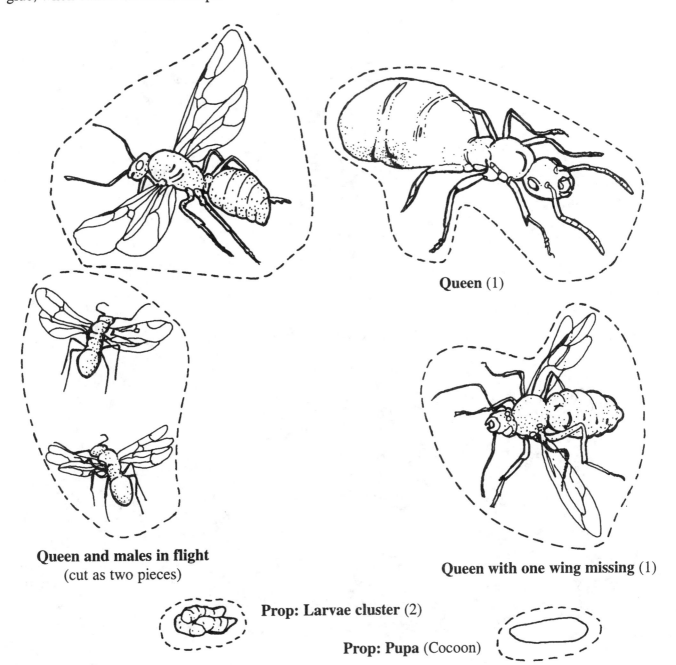

Queen (1)

Queen and males in flight
(cut as two pieces)

Queen with one wing missing (1)

Prop: Larvae cluster (2)

Prop: Pupa (Cocoon)

Life in an Ant Colony *(cont.)*

Ant Puppets and Props *(cont.)*

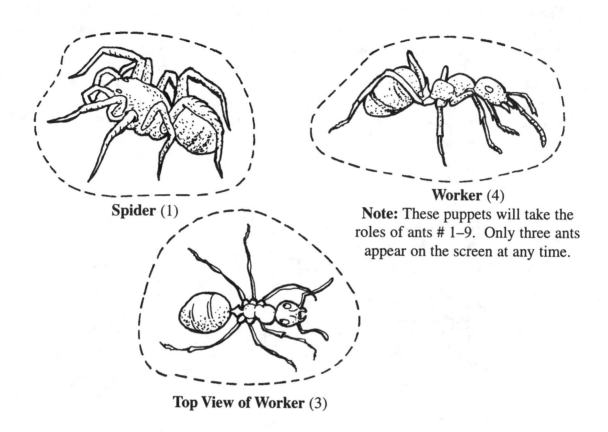

Spider (1)

Worker (4)
Note: These puppets will take the roles of ants # 1–9. Only three ants appear on the screen at any time.

Top View of Worker (3)

Queen's Chamber Scenery

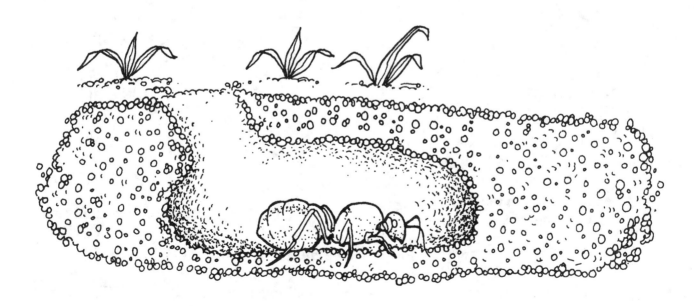

18

Life in an Ant Colony *(cont.)*

Ant Colony Scenery

To the Teacher: Make a transparency of this picture to serve as background scenery for the puppet show.

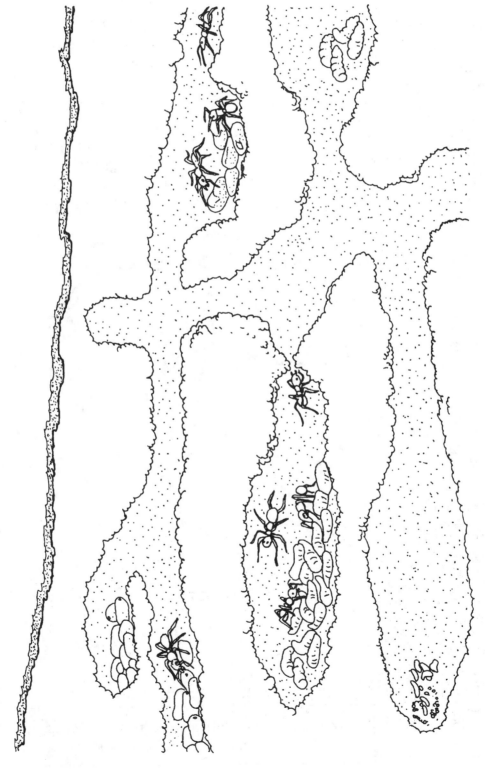

Life in an Ant Colony *(cont.)*

Ant Colony Cover Page for Scenery

To the Teacher: Make a copy of this page and cut it as indicated. Tape this page on top of the transparency of the ant colony scenery. Fold back each section as called for in the script to reveal an area of the colony.

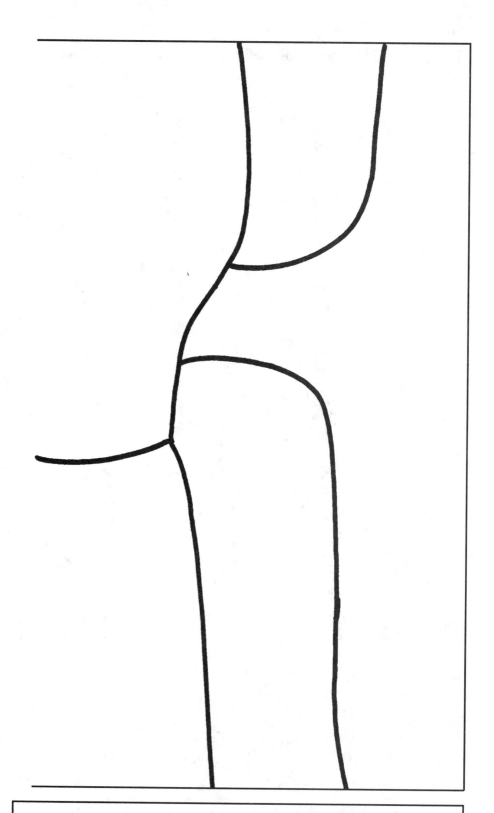

Directions: Tape each edge to the transparency on the preivous page. Make cuts along the thick lines when called for in the script.

Life in an Ant Colony *(cont.)*

Top View of Ground Near Colony Entrance

To the Teacher: Make a transparency of this picture to use as scenery for the shadow puppet show when called for in the script.

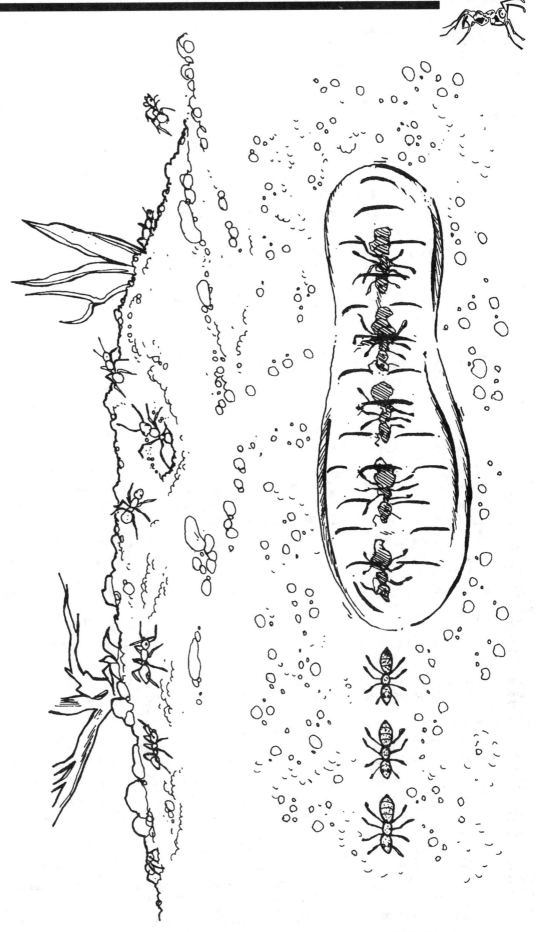

Inside An Ant Farm

Topic: Life in an ant colony

Objective: Students will learn the changes over time in an ant colony by observing a commercial ant farm.

Materials:

- 1 or 2 *commercial ant farms with live ants and instructions for assembling an ant farm
- Ant Record (made during earlier study of wild ants)
- Optional: video camera

Insect Lore (see Resources section) supplies ant farms of various sizes that include an ant watcher's manual, sand, and a coupon to return for a supply of live ants.

Caution: Do not add wild ants to those already in the ant farm; they will kill each other.

Preparation:

- Order ant farms early enough to allow time to send in the coupon for the live specimens.
- Assemble the ant farm(s) and label one side A and the other B.

Procedure:

- Discuss the Ant Records students made during their observations of wild ants. Review what they learned from the puppet show script.
- Arrange the students so they will be able to see the ant farms. Tell them that the ant farms will simulate the ant colonies which are underground and thus not visible to them.
- Explain that they will be able to observe what goes on beneath the ground in an ant colony by watching the ant farms over the next several weeks.
- Show them the live ants which were supplied for the farm. Explain that these are all worker ants; no queen ant is included. Let students look at the ants to compare them with the wild ants they observed in an earlier lesson.
- Place the ants and food into the farm(s) according to the distributor's instructions.

For Discussion:

- Let the students begin to observe and discuss what they see happening as the ants begin to investigate their new home.
- Point out that each side of the ant farm has been labeled A or B. Explain that they will be making daily observations to draw the chambers they see appearing in the colony.

Follow Up:

- Students should record the changes they see in the ant colony each day on blank white paper, one piece for each side. If two ant farms are used, divide the observation teams between them.
- Schedule students at the ant farms for a 5–10 minute period each day. Allow only four observers at each ant farm during the period. After the ants have begun to build chambers, cover one side of the farm for several days with black paper. Have students observe and draw the changes when it is removed.
- A time-lapse video record of the changes in one ant farm can be done by recording about 30 seconds of tape daily. Mark the location of the ant farm and the camera position so they will be in the same spot every time. Always tape the exact same section of the ant farm and take some close-up shots to show details. The final video will show the changes as a continuous sequence. This will enhance student understanding and observations.

An Ant's Story

To the Teacher: This open-ended question will assess the depth of understanding students have gained through the activities in the study of ants which they have just completed.

Name _____ **Date:** _____

To the Student: Pretend you have a dream in which you are changed into an ant. You may choose to be a worker or a queen ant. Tell what your life is like in the colony and then draw pictures to go with the story. Your story and pictures should be as detailed as possible to show what you have learned in this study of ants.

_____
_____
_____
_____
_____
_____
_____
_____

Picture of me as an ant

Picture of my colony in action

Being an Inventor

Teacher Background: Inventors have to be creative, patient, and good problem solvers. Their ideas for new inventions may spring from many different experiences. Nearly everything we use was developed by people who were willing to devote their time, energy, and efforts to turn their ideas into reality. Without inventors, we would not have such wonderful things as the light bulb, airplanes, computers, microscopes, telescopes, and millions of other things which add to our quality of life.

Students are introduced to the thrill of being an inventor after learning how to work with batteries, wires, and small light bulbs. This safe and easy way to use electricity will enable the students to create uses beyond those they learn about through the introductory lessons. They first begin by discovering how to make a bulb light using only a wire and battery. The next step leads to investigating how the light bulb works. Students invent a simple flashlight and then extend their understanding of electricity by creating electric circuits.

The final assessment for this unit is performance based since it consists of a variety of activities directly related to those students have done throughout this study.

A logical extender for this unit is for students to investigate the lives of famous inventors or developers such as Edison, the Wright brothers, Steven Jobs, and Stephen Wozniak. They will find that many inventions are actually refinements of ideas which originated earlier. This may give the students the incentive to become inventors themselves. They can be encouraged to think of something that they see a need for in their own lives and design an invention to fulfill that need.

Applying their knowledge of electricity, using batteries and a small motor to invent a new toy or tool, would be one way to begin.

The teacher may want to organize an *Inventor's Fair* to provide students with the incentive for designing an invention to exhibit for others to see. These designs may be drawings, scale models, or working models of the inventions. It could become a lifelong hobby with some of the students once they see how exciting it is to be an inventor. Perhaps one of your students may invent something in the future which will be of great use to people everywhere. What a great reward that would be, to know it all began when you awakened his/her interest in inventing.

Can You Light the Bulb?

Topic: Using electricity created by batteries

Objective: Students will light a 2.5 volt bulb, using only a wire and D-sized battery.

Materials: (Each student will need the materials listed below.)

- D battery
- 2.5 volts light bulb
- *6" (15 cm) of insulated wire with both ends stripped of wire

*Use telephone wire purchased at a building supply store.

Motivator:

- Ask the students if they have ever used a flashlight.
- Ask them to make drawings to show how they think one works.
- Collect their drawings to use later in this unit when they invent their own flashlights.

Procedure:

- Explain that they will each receive a light bulb, one wire, and a battery to connect in different ways to light the bulb.
- Distribute the battery, wire, and light bulb to each student. Do not give them any assistance as they work to get their light to work. Tell them to try to figure out how to do this on their own.
- When the first student is able to light his/her bulb, ask him/her not to show others how he/she did it. Explain that you will have students draw how they did it on the board once a few others have discovered how to get their bulbs to light.
- After more students are successful, let the first student draw how he/she did it on the board. Draw a large battery for him/her so he/she will know that you expect the picture to be big enough for all to see the details of how the wire connects the battery and bulb to make it light up.
- Let the students know that there are four ways to get the bulb to light.
- Have the students get their bulbs to light in four different ways and make drawings of each method.

For Discussion:

Have students show pictures of the four different methods they used to light their bulbs.

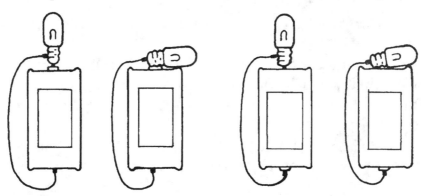

How Does A Light Bulb Work?

Topic: Components of an electric light bulb

Objective: Light bulbs will be investigated to discover what makes them work.

Materials: Each student will need the following:
- same materials as in the previous activity: *Can You Light the Bulb?*
- magnifier
- two 3" x 5" (8 cm x 13 cm) file cards and 6" (15 cm) of yarn

Additional materials for entire class use:
- burned out 2.5 volts bulb(s) and variety of clear light bulbs
- transparency of light bulb picture (page 27)

Preparation:

Carefully remove the metal casing from a 2.5 volts bulb, using wire strippers or other tools. Wear gloves so you can hold the glass bulb as you work. Try to expose the wires which lead from the glass bulb into the base, but preserve the connections between the wires and the base.

Motivator:
- Have students use the wire, battery, and bulb to light up their bulbs in the four different ways they discovered in the first lesson.
- Tell them to use one of these methods again and this time look closely at the light bulb to see what part is glowing in the bulb (filament).

Procedure:
- Have students draw a picture of the light bulb on the file card, large enough to fill the card.
- Tell them to use their magnifiers to examine the bulb inside and out so their drawings will show every detail they see.
- It is especially important that they notice where the wires from inside the bulb connect to the bottom and side of the metal casing of the bulb.

For Discussion:
- Show students the transparency of the light bulb picture and explain its parts. Have them compare this drawing with their own.
- Show them the bulb(s) with the base removed.
- Have students reconnect their light bulb to the battery. Tell them to look closely at the contacts between bulb, wire, and battery.
- Ask them if they can now see why one end of the wire contacts the battery and the other end can only touch the bulb while the bulb contacts the battery.
- Have students use another file card to draw a battery, and use the yarn as wire to show the points of contact necessary to make it light.
- Show students the examples of clear light bulbs to examine. They will find they are constructed much like the smaller ones they have been using.

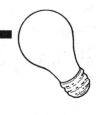

How Does a Light Bulb Work? *(cont.)*

Parts of a Light Bulb

2.5 volt bulb

100 watt bulb

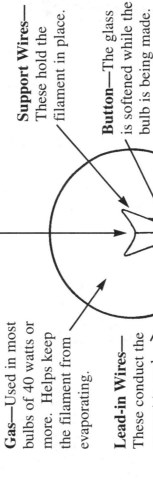

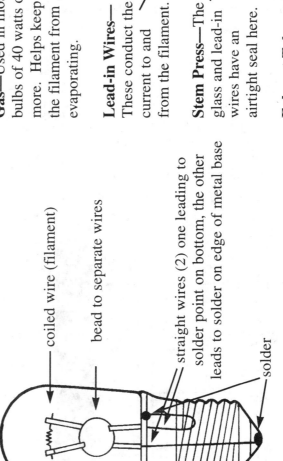

Filament—The wire that heats up to *incandescence* (in-can-des-cents), or glowing, white heat. Edison used a cotton sewing thread burned to an ash as his first filament. Today, the filament is made of *tungsten*, which does not melt except under extreme heat.

Gas—Used in most bulbs of 40 watts or more. Helps keep the filament from evaporating.

Lead-in Wires—These conduct the current to and from the filament.

Stem Press—The glass and lead-in wires have an airtight seal here.

Exhaust Tube—Through this tube air is taken out of the bulb and the gas is pumped in. The tube is sealed off so the base will fit over it.

Support Wires—These hold the filament in place.

Button—The glass is softened while the bulb is being made. The support wires are stuck in it.

Heat Disc—Used in bulbs of high wattage. Protects the bottom of the bulb from heat.

Fuse—This helps keep the bulb from cracking and prevents blowing of electric fuses.

Base

coiled wire (filament)

bead to separate wires

straight wires (2) one leading to solder point on bottom, the other leads to solder on edge of metal base

solder

Predicting Success

Topic: Investigating various ways to light a bulb

Objective: Students will identify correct methods of connecting a wire, bulb, and battery to make the light work.

Materials:

- same materials as in the previous activity: *Can You Light the Bulb?*
- transparency and copy of Will the Bulb Light? worksheet (page 29) for each student

Procedure:

- Distribute the worksheet to the students and review the instructions, using the transparency.
- When all students have finished the worksheet, have them check their answers with the bulb, battery, and wire.
- Have each student draw another way to connect the bulb to the battery so it will light up.

For Discussion:

- Have students exchange papers to see if they can use their bulbs, batteries, and wires to get the light to work, as shown in box #8.
- Let students demonstrate some of the different ways this can be done.

Follow Up:

Show students the diagram of the parts of a battery.

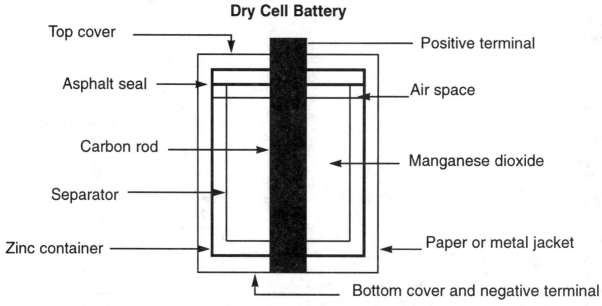

A dry cell battery consists of a zinc container with chemicals inside it that produce an electric current by reacting with one another. The battery has a positive (+) and negative (-) charge. The container is the negative terminal, a carbon rod in the center serves as the positive terminal.

Predicting Success *(cont.)*

Will the Bulb Light?

To the Student: Some of the drawings below of a battery, wire, and bulb connected in different ways will light the bulb; others will not. Examine each drawing carefully and then circle **Y** if the bulb should light and **N** if you think it will not work.

- Use a wire, bulb, and battery to test all of these pictures and check your answers.
- Draw the changes needed in the pictures marked N to show how to make a bulb light.
- In the box, draw another way to connect the battery, bulb, and wire to make the bulb light up.

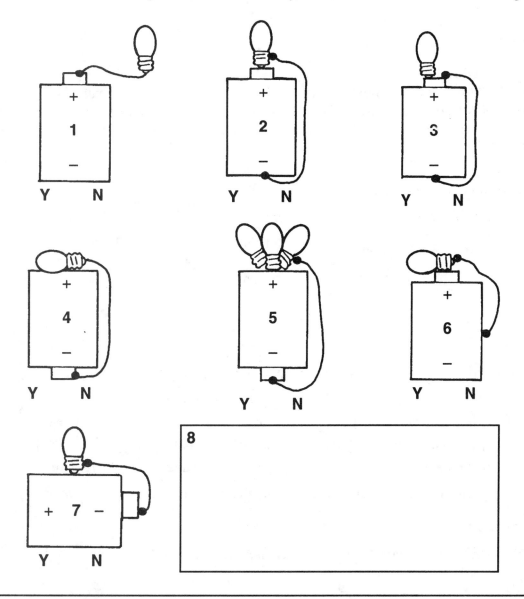

- Use a wire, bulb, and battery to test all of these pictures and check your answers.
- Draw the changes needed in the pictures you marked **N** to show how to make the bulb light.
- In the box, draw another way to connect the battery, bulb, and wire to make the bulb light.

Inventing a Flashlight

Topic: Creating a tool using a bulb, battery, and wire

Objective: Students will apply their knowledge of electricity to make a simple flashlight.

Material:

- same materials as used in *Can You Light the Bulb?*
- cardboard tubes (toilet paper rolls will do)
- masking tape
- pieces of aluminum foil approximately 2" (5 cm) square

Preparation:

- Construct a flashlight from a bulb, battery, and one long wire. Use the cardboard tube to hold the batteries. Use tape and foil to hold the bulb on the battery and make contact with the wire so it will light. The foil can be used as a reflector around the light and as a switch. The drawing below shows one way to make a flashlight; there are other ways of constructing it.

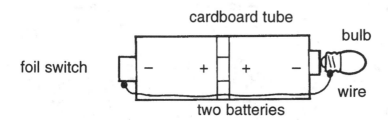

Procedure:

- Divide the students into pairs and give each pair a bulb, long wire, and two batteries. Have them put these together to light the bulb.
- Distribute the cardboard tube, piece of foil, and a 6" (12 cm) piece of tape to each pair.
- Explain that you want them to use these materials to create a flashlight that can be turned off and on. Tell them that they may have more tape or foil if needed but not more wire.
- Monitor and encourage the students as they work but let them create their own designs.

For Discussion:

Have the students share their flashlight designs, including making a drawing of it on the board.

Follow Up:

Darken the room and have students see how far their light travels. Let them compare the distances and see if the design of the flashlight determines how far the light travels.

Creating Electric Circuits

Topic: Constructing a simple electric circuit

Objective: Students will make an electric circuit using a battery power.

Materials: Each student will need the following:

- 2 wires
- battery
- light bulb

- *light bulb socket
- *battery holder

*See the Resources section to locate a source for this equipment.

Preparation:

- Order the bulb sockets and battery holders in time to be used for this lesson.
- Experiment with connecting these in order to experience some problems students may have.

Procedure:

- Distribute the materials to each student and tell them they are to experiment with assembling them to light the bulb.
- Show them how to use the connectors on the sockets as shown below, but do not show them how to assemble the circuit.

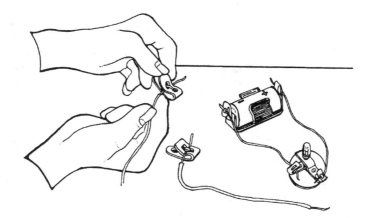

For Discussion:

- Have the students discuss how the electricity transfers to the light bulb through the socket. (They should notice that there is a strip of metal beneath the bulb and another around the bulb, thus making contact on the bottom and side of the bulb.)
- Let students discuss how this compares to lighting the bulb with only a battery and one wire. (The same contact points on the side and bottom of the bulb are still being made in the socket as was required, using only the bulb and one wire with the battery.)

Follow Up:

When each student has successfully connected his/her equipment to light the bulb, let each experiment with another student to see if they can light up both bulbs with two batteries. Have each pair show what they did and see how many variations the students discovered.

How Does a Circuit Work?

Topic: Electrical circuits in series and parallel

Objective: Students will construct circuits to discover the series and parallel methods of connecting the wires.

Materials: Each student will need the following:

- battery, battery holder, bulb, socket, 2 wires
- Constructing Circuits worksheet (page 33)

Preparation:

Follow the instructions on the worksheet to construct the circuits and then conduct the experiments. You will find that when one bulb in the parallel circuit is loosened in its socket, the other lights continue to burn. When this is repeated in the series circuit, all bulbs go out when any of the others are loosened. The electric current can flow around the loosened bulb in a parallel circuit through the wires and into the next bulbs. The series circuit has only a single wire linking the bulbs, and, therefore, when one bulb is loosened, electricity cannot flow beyond that point regardless of which bulb is out.

Procedure:

- Divide the students into groups of three or four.
- Distribute the materials to each student and let each one construct a circuit.
- Let students experiment with the equipment, linking up bulbs, batteries, and sockets in various ways. Tell them they should never link up more than two batteries to a single bulb or it will burn out.[1]
- Distribute the work sheet and have them construct the parallel and series circuits and then answer the question.

For Discussion:

Have students discuss the answers to the work sheet.

---

[1] The bulbs are only 2.5 volts and each battery is 1.5 volts; thus, more than two batteries will burn out the bulb. If a bulb does burn out, have students examine it with a magnifier to see the broken filament that burns through and breaks the flow of electricity.

How Does a Circuit Work? *(cont.)*

Constructing Circuits

To the Student: *Parallel* circuits are constructed with two wires running side by side. *Series* circuits have a continuous circle. Construct each of these circuits as shown below. Then answer the questions as you conduct an experiment with these circuits.

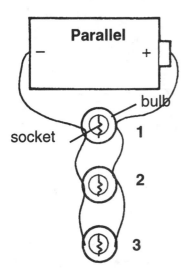

1. Unscrew the bulb in socket 1 just until it goes out. What happens to the bulbs in the other sockets?_____

2. Tighten the bulb again and then repeat this experiment by unscrewing one bulb at a time. Did the other bulbs go out? _____

3. Use a colored pencil and trace the electricity from the battery to bulbs 1, 2, and 3 and then back to the battery.

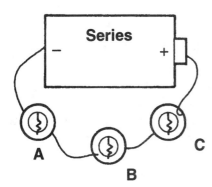

4. Unscrew the bulb in socket A. What happened to the other bulbs? _____

5. Repeat this by unscrewing one bulb at a time. What happened to the bulbs?_____

6. Use a colored pencil and trace the electricity from the battery to bulbs A, B, and C.

7. Compare the parallel and series circuits. Tell what was different about them when you unscrewed the bulbs. _____

_____

8. Explain why they were different. _____

_____

Inventing Circuits

Topic: Electric circuits

Objective: Students will apply what they have learned about circuits and develop combinations of them to increase their understanding of how they work.

Materials:

copy of the worksheet Will the Circuit Work? (page 35) for each student

Each Group Will Need the Following:

- 8 wires
- 2 batteries
- 2 battery holders
- 4 sockets
- 4 bulbs

Procedure:

- Give students worksheets and have them predict if the circuits will or will not work.
- Distribute the materials to each student and then divide students into groups of three.
- Let each group use their equipment to see how to find a way to make the circuits work which would not light the bulbs.

For Discussion:

Have students compare their worksheets to show their answers and how they corrected the circuits that would not work. Check them for accuracy.

Follow Up:

- After students have completed the worksheet, have them do the "Challenge" by inventing their own circuits. They should draw them on the backs of the worksheets and identify them as parallel or series circuits.
- Have each group show their circuit invention.

Inventing Circuits *(cont.)*

Will the Circuit Work?

To the Students: Look at the circuits below carefully to decide if you think the light bulbs will light up or not.

1. Circle **Y** if you think the circuit will work or **N** if you think it will not.

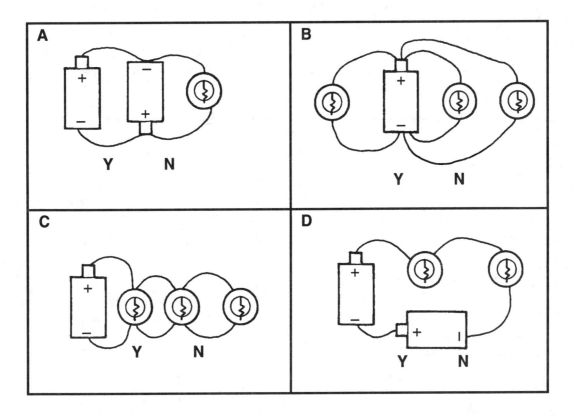

2. Construct each circuit and check your answers.

3. Write the letter(s) of those circuit(s) where your prediction was wrong. _____

4. Which circuit did not work? _____ Why? _____

_____

Work with the batteries and light bulb to make the circuit work. Describe what you had to do to get the circuit to work. _____

_____

_____

Inventing Circuits *(cont.)*

Will the Circuit Work? *(cont.)*

5. Find the circuits which are parallel and write their letter(s)._____

6. Find the circuits which are series and write their letter(s)._____

7. Check your answers for #5 and #6 by constructing the circuits and unscrewing one bulb. Remember that in a series, if one bulb is out, none of the others will light. If the circuit is parallel, all bulbs will still light up even if one is out.

Challenge: Use the batteries, wires, sockets, and bulbs to invent a circuit which you have not ever built. Draw the circuit you invented below in the box and tell if it is a parallel or series, or a combination of these.

Searching for Conductors

Topic: Conductors of electricity

Objective: Students will test a variety of objects to find what will conduct electricity.

Materials: Each group will need the following:

- battery
- battery holder
- 3 wires
- socket
- light bulb

- variety of materials to check as conductors, including metal and non-metal items such as these: coin, pipe cleaner, marble, foil, paper clip, and *pencil with lead exposed
- baggies for materials being checked

*Pencil lead is graphite and will conduct electricity.

Preparation:

Assemble materials which students will check as conductors and place them into baggies. Each group should have the same items to check.

Procedure:

- Divide the students into groups of 3 or 4 and distribute a baggie of materials to each of them. Have them sort the materials into piles of things which they think electricity will pass through and those which they don't think will conduct electricity.
- After the sorting is complete, list the items on the board in the two categories, according to the students' ideas. There may be disagreements so some items may be in both lists. This should be permitted since it will be corrected after all items are tested.
- Distribute the electrical equipment to each group and show them how to construct a conductor tester (see diagram) that will be used to test each item for conductivity.

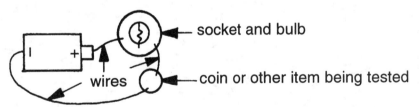

- Demonstrate how to touch the wires leading from the battery and the socket to the item being tested, without the wires touching each other. Explain that if the light goes on, electricity has passed through the item, proving it is a conductor.
- Let students test all the items they have just sorted.

For Discussion:

- Have students discuss what they found and correct their list of conductors.
- Ask them to see what all the conductors have in common. (They are all metal.)

Follow Up:

Have students test a variety of things in the room to find conductors.

Mystery Connections

Topic: Wiring electric connections

Objective: Students will be presented with a variety of boxes that have been wired and diagram the connections.

Materials:

- 8 shoebox lids
- various lengths of .5" (1.5 cm) foil strips or insulated wire
- brass brads
- tagboard
- conductor testers used in last activity (page 37)
- Mystery Connections Record (page 41)

Preparation:

- Label the eight box lids A–H and then number them as shown in the diagrams.
- Punch holes near the numbers and insert brads through them, keeping the heads of the brads on the outside of the lid.
- Attach wires or thin strips of foil to the brads on the undersides of the lids. Be sure the contact is firm by bending the prongs of the brads over the foil or wrapping the tip of the wire around them. If foil is used, place masking tape over each strip, including the prongs of the brads. The tape is an insulator to prevent the electricity from passing between the foil strips where they cross. Follow the pattern shown in the diagrams to connect the wires or foil.
- Write the numbers near the prongs inside the lids to correspond with those on the outside.

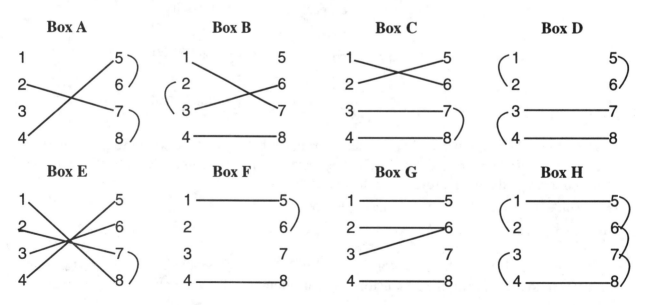

- Cut oaktag so it will fit inside the box lids and cover the wires. It will hide the wires and will be removed for students to check their work so do not tape it into the lids.

Mystery Connections *(cont.)*

Procedure:

- Divide the students into eight groups and distribute a conductor tester, box lid, and Mystery Connections Record to each of them.

- Read the instructions with the students and have each group test the connections between #1 and #2 brads. Have them continue the work and monitor their progress.

- When all brads have been tested, students remove the tagboard which covers the wires in the box and draw where the wires are, using the record in the "Wires" column of their record sheet.

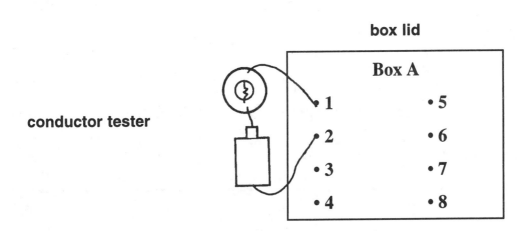

For Discussion:

- Have students compare the location of the wires with the connections their tester detected. Ask them to explain how the light could go on without having a wire connected directly from that brad to the one which was touched to turn the light on. (Electricity will flow via another connection, as shown in the diagram below.)

| 1   5 | The light in the tester will work when brads 2 and 5 are touched because electricity will be able to flow from 2–6–5 to close the circuit. |
| 2————6 | |

Mystery Connections *(cont.)*

Follow Up:

Let students test more of the boxes to find the various methods of attaching the wires.

Electronic Question and Answer Board

1 •	A •
2 •	B •
3 •	C •
4 •	D •

Answer Sheet

Problems		Answers (Letter)
1.	4 x 5 =	_____
2.	2 x 3 - 2 =	_____
3.	16 ÷ 4 =	_____
4.	3 x 4 ÷ 2 =	_____

Mystery Connections *(cont.)*

Mystery Connections Record

To the Students: You will use your conductor tester to find the connections between the brads in the shoebox lid. Follow the instructions below to see how to do the test.

1. Write the letter of your box on the record below in both the columns.

2. Press the tip of one wire of the tester to brad #1 and press the other wire to brad #2.

3. If the light goes on, draw a line between #1 and #2 on the record in the "Connections" column. Do not draw a line if the light does not go on.

4. Continue to press one wire to #1 brad and move the other wire to the #3 brad. Again, draw a line on the record if the light goes on.

5. Repeat this process, always holding the wire on #1 brad until you have checked the brads 2–8 and recorded the results.

6. Next, move the wire to brad #2 and test all the others to find the connections, recording those which light the bulb.

Connections Box _____		Wires Box _____	
1	5	1	5
2	6	2	6
3	7	3	7
4	8	4	8

7. After all the brads have been tested for connections, remove the cover inside the box and look at where the wires are. Draw these wires on the record in the "Wires" column.

8. Compare the location of the wires and the connections you recorded. Sometimes a brad did not have a direct wire connected to another brad but the light went on when you tested them. Draw and explain how the electricity could flow between these unconnected brads.

9. Exchange boxes with another group and test to find the connections. Record these below.

Connections Box _____		Wires Box _____	
1	5	1	5
2	6	2	6
3	7	3	7
4	8	4	8

Inventing Mystery Connections

Topic: Electric circuits

Objective: Students will wire their own mystery connections to learn more about the paths of electricity.

Materials:

- heavy cardboard about 6"x 8" (15 cm x 20 cm)
- conductor tester
- brads
- wires or ½ inch (1.5 cm) foil strips
- masking tape for each group

Preparation:

Drill holes in the heavy cardboard to make double rows of four holes.

Procedure:

- Review the Mystery Connections activity and show one of the box lids to remind students of how they were wired. Tell them they will work in groups and invent mystery connections on a card for another group to test.
- Divide students into groups of three or four and then distribute one cardboard, eight brads, and eight wires or foil strips to each group. Explain how to apply the wires or foil. If foil is used, demonstrate how to cover it with tape to secure it to the brads and insulate it from other wires.
- Have students number the holes on both sides of the card.
- Distribute the worksheet to each group and then let them begin inventing their mystery card.
- Distribute the materials for the conductor tester to each group as they finish their connections so the students can test them and be sure they match their answer sheets.
- Have each group tape their card to their table so the connections are hidden from sight.
- Move the groups to new stations so they can use their testers on different mystery cards. Have them record the results on their worksheets and then take the card off the table to find the actual connections.

For Discussion:

Have each group discuss what they discovered when they invented their mystery card, as well as when they tested another group's card.

Follow Up:

- Students can turn their connection cards into a question/answer board by connecting pairs of brads. Have one row of four brads numbered and label the other row A–B. Make a list of four questions and four answers; these may be in any subject area (i.e., math, science).
- Number the questions and letter the answers. Be sure the answers are in a different order from the questions.
- Connect the brads to match a number with the letter which is the correct answer. Cover the back of the card with another card to hide the connections.
- Exchange cards and the questions/answers list with another group.

Inventing Mystery Connections *(cont.)*

To the Students: Now it is your turn to invent a mystery connections card. You will need the following materials:

- cardboard with eight holes drilled in it
- 8 wires or foil strips
- 8 brass brads
- roll of masking tape

Instructions:

- Neatly write the number of each hole near it to match the diagram.
- Place a brad through each hole and spread out the prongs.
- Number the dots, which are brads, on the diagram of your mystery card below.

Mystery Card Connections

```
    1        2        3        4
    •        •        •        •


    •        •        •        •
    5        6        7        8
```

- Draw where the wires or foil will be placed to connect the brads. Not all brads need to be connected. Think of ways to make the mystery card connections challenging since another group will be testing it.
- Connect the brads with wire or foil. Be sure to make tight contact between the prongs of the brads and the wire or foil.
- Write the numbers of brads which should form circuits and make the tester light work. Fill in the information on the chart below for the circuits you have created, which may be fewer than four.

Circuit Brad Numbers _____   **Circuit Brad Numbers** _____

Circuit Brad Numbers _____   **Circuit Brad Numbers** _____

- Use the conductor tester and see if the brads you listed for each circuit make the light work.

Inventing Mystery Connections *(cont.)*

Testing a Mystery Card

To the Students: Now it is your turn to test the mystery card invented by another group. Use the conductor tester and draw lines between the brads which make the light work.

A

1	2	3	4
•	•	•	•
•	•	•	•
5	6	7	8

Remove the paper from the mystery card and draw the connections you see on the card.

B

1	2	3	4
•	•	•	•
•	•	•	•
5	6	7	8

Did you find any two brads which were not directly connected to each other but the light worked when they were both touched? If so, list these pairs below.

Pairs of brads which were not connected but made the light work:

_____   _____   _____   _____   _____   _____

On diagram A draw dotted lines to show how the electricity flowed between these pairs of brads even though there was no direct connection between them.

Electric Circuits

To the Teacher: This is a performance-based assessment which will enable students to demonstrate what they have learned from this study. There will be four tasks for students to perform independently. The activities, materials, and answers are described on the chart.

Task	Activity	Materials	Answers
1	lighting a bulb in four different ways	• 1 bulb • 1 D-cell battery • 1 long insulated wire	
2	test for conductors	• bulb, battery, socket, 3 wires • items used in activity: "Searching for Conductors" (page 37)	• list of items with metal in them and pencil lead (graphite) • sentence stating that only metal conducts electricity
3	identifying circuits	• Box C, used in "Mystery Connections" (page 38)	• connections: 1–6; 5–2; 3–7; 7–8; 4–8; 3–4; 4–7; 8–3 • electricity travels between 3 and 4 via a route through 3–7–8–4.
4	predicting circuits	• answer sheet	• NO: #1 (needs another wire); #4 (connect wire to socket); #5 (add bulb)

Administering the assessment:

- Gather the materials needed for each of the tasks. This assessment can be done by the entire class at one time by providing enough stations for each student to work independently. Divide the work areas with cardboard barriers to provide privacy. Work out a rotation system for students to move from task to task on a given signal. The diagram above shows one possible method to do this. Each station has the materials for that task, except #4 which requires only the answer sheet. Label each station with the task number.
- Students should be given a set time to complete the task and then all rotate at the same time. Adjust the time to the ability levels of the students.
- During the assessment, students should not be permitted to talk.
- Monitor the students' progress and be prepared to substitute equipment and pencils as needed.

Instructions to be read to the students:

- Review the tasks students will perform during the assessment and explain the rotation system.
- Tell the students you will announce when there are five and then two minutes remaining for them to work on the task.
- Let students know that there is to be no talking or looking at another student's work.
- Explain that when they complete the task, they should clean up the station so it looks like it did before they began work. Tell them to let you know if they need any materials as they work, including a sharpened pencil.
- Be sure students find the numbers of their stations on the answer sheets before beginning.

Electric Circuits *(cont.)*

Task 1

Connect the wire, battery, and bulb to light the bulb. Draw a diagram to show how you did this and label the parts of your drawing. Be sure to use large dots to show where the connections are between the parts.

Now, find three more ways to connect the parts to make the bulb light and draw them.

Task 2

Construct a conductor tester from the materials at this table and test the items for conductivity. Draw what your conductor tester looks like.

Conductor Tester

List the materials which conducted electricity and made the bulb work.

_____ _____
_____ _____
_____ _____
_____ _____

Write a sentence which tells what you found about conductors of electricity.

_____

_____

Give an example of something else which would conduct electricity but is not with the materials at this table. _____

Electric Circuits (cont.)

Task 3

Construct a conductor tester from the materials on the table and then check to find where the connections are between the brads in this box lid. Record your test in the box on the left.

Connect the numbers to show which made the light of the tester work.

1	5
2	6
3	7
4	8

Remove the cover from the box lid. Draw where the wires are connected to the brads.

1	5
2	6
3	7
4	8

Which brads made the tester light work but were not connected by wire?_____

Explain how the electricity could go between these brads even though they did not have a wire connecting them. _____

Task 4

Look at the circuits shown below and decide if they will work. If they will, circle **Y**; if not, circle **N**. Draw the changes needed to make the circuits work where you circled **N**.

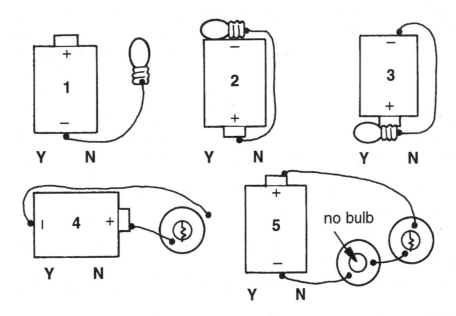

Being a Chemist

Teacher Background: Most students think that the study of chemistry is about explosions, smelly stuff, and magical potions. Chemistry is actually the study of matter and the way it changes, sometimes with surprising or spectacular results. This series of lessons is designed to simulate how chemists are able to identify different types of matter. In this case, students will work with indicator dyes to be able to identify matter as an acid or a base. This will appear to be magic since the indicator dyes undergo surprising color changes when mixed with the acids or bases.

Some indicator dyes will only react to acids, others to bases (also called alkalies), and some react to both. Extracts from some vegetables such as red cabbage or colored flower petals can be used as indicator dyes. Other acid/base indicators include bromothymol blue (BTB) and phenolphthalein (fe´ nol thal´ en), and pH paper which are available commercially. The pH scale ranges from 1–14. The pH paper is actually thin filter paper which has been impregnated with an indicator dye. Like thermometers, pH paper has been calibrated to provide a value. A pH paper ranging from 1 to 12 is used for these activities.

The teacher initiates this series of lessons by performing a *magic show,* using indicator dyes. The activities culminate with a simulation of students acting as scientists helping to discover the cause of pollution in an imaginary lake. After finding the cause of the problem, students do an experiment to neutralize an acidic substance.

Safety Note: These lessons should be done in a well ventilated room or outdoors. Caution the students in the use of ammonia and vinegar. Both have strong fumes and can burn if they get into the eyes. It is wise for students to wear safety goggles and wash their hands after this activity.

The risk of using vinegar and ammonia is greatly reduced by using small quantities of the liquids in these activities. It is suggested that 3.5 mL microscale pipettes be used rather than droppers. Pipettes should be designated for one chemical only (e.g., ammonia) to avoid contaminating the test samples. These can be ordered in packages of 20 from Flinn Scientific, Inc. (see Resources). You will need four packages for all of these activities.

Reaction strips, which are plastic strips of tiny cups, are recommended for holding the indicator dyes. Order reaction strips with 12 wells and 0.3 mL capacity. Since they can be washed and reused, you should be able to do all the lessons with 10 of these strips. These are fairly inexpensive and are also available from Flinn.

Chemical Magic

Topic: Using indicator dyes

Objective: Students will use simple, safe indicator dyes to learn how to identify acids and bases.

Materials:

- *safety goggles for students (often available in paint and building supply stores)
- 1 qt. (1 liter) red cabbage extract (see recipe on page 51)
- *1 gram bromothymol blue (BTB) powder
- uncoated laxative tablets (from drug store) for phenolphthalein
- *2 pH paper vials of 100 strips each with color code (1–12 range)
- *27 thin stem microscale pipettes (ends may be cut to shorter length)
- *microscale 8-well reaction strips (optional: 1 oz [30 mL] clear plastic cups)
- *5 test tubes (about 20 mL capacity)
- 6 Styrofoam cups (or a test tube rack)
- 1 qt. (1 liter) white vinegar
- 1 qt. (1 liter) plain ammonia (without soap added)
- Testing Indicator Dyes record sheet (page 52)
- 8 trays (used to deliver materials to students and as a work surface in the event of spills)
- 1 qt. (1 liter) distilled water
- 8 pairs of scissors

*These materials may all be ordered from Flinn Scientific, Inc. (see Resource section).

Preparation:

- Follow the directions on page 51 to create the indicator dyes needed for this lesson.
- Make 8 copies of the pH color code and seal them in clear plastic wrap or laminate them.
- Make small adhesive labels for the 16 pipettes as follows: vinegar (8 pipettes), ammonia (8), water (8). Fill the pipettes about 1/2 full with vinegar, distilled water, or ammonia.
- Use another set of three pipettes and mark them V, A, and W with small letters in permanent ink.
- Fill these pipettes with vinegar (V), ammonia (A), and distilled water (W).
- Make holders from the six Styrofoam cups by turning them upside down and punching a hole in the bottom of each so it will hold one test tube snugly. (Optional: Use a test tube rack.)
- Fill the test tubes 1/2 full of (3) cabbage extract, (1) phenolphthalein, (1) BTB.
- Fill five of the reaction strip cups half full as follows: (3) cabbage extract, (1) phenolphthalein, (1) BTB.
- Place the trays, pipettes of vinegar, filled reaction strips, and record sheets on a table.

Motivator:

- Present a chemical magic show to introduce this lesson. Practice this magic act before doing it for the students to perfect the surprise aspect of the demonstration.
- Place the test tubes containing the cabbage extract, BTB, and phenolphthalein where students can see them. Point out the color of each of the three indicators but do not tell what they are.
- Show that the liquid in pipettes marked V, W, and A are clear. (Do not show the letters.)

Chemical Magic *(cont.)*

- Tell students that you are about to do a magic show. Explain that they should watch closely to see what happens to the liquids in the test tubes when you put a few drops of the clear liquid into them.

- Hold up one of the test tubes with cabbage extract in it and drip some vinegar into it. Shake the test tube until the extract changes to violet. (Add more drops if no reaction occurs.)

- Repeat with ammonia in another test tube of cabbage extract (color becomes green).

- Add water to the third cabbage juice test tube (no color change).

- Next, use the phenolphthalein and place a few drops of vinegar into it (no color change). Use the same test tube to add a few drops of water (no color change) and then ammonia (bright bursts of pink color are seen and then disappear). Add more water until the dye remains pink.

- Last, use the BTB sample and drop in vinegar (color changes to orange/yellow) and then add drops of water (no color change) and ammonia (color returns to blue).

- Take a piece of pH paper and show its color. Cut it into three pieces. Add a drop of vinegar to one piece, let students observe the color change. Add a drop of water to the second piece (no color change) and a drop of ammonia to the other piece (color change, different than with vinegar).

- Ask the students to describe what they saw happen. Tell them that although this looks like magic, it is actually chemistry and that they will now have the chance to do the magic.

Procedure :

- Divide the students into eight groups. Distribute goggles to each student. If goggles are limited, give one pair to each group and have the students take turns completing the tests.

- Tell each group to send one person to the supply table to take a tray and pick up one pipette of vinegar, a 6-well reaction strip, and one record sheet.

- Have the students place their indicator dyes in the box on the record sheet. Assist them as they identify the dyes by writing the names on the paper in front of the cups.

- Guide them as they conduct the first test, using vinegar and the cabbage extract and record the results. If they are ready, let them complete the remaining tests on their own.

- As each group completes the vinegar test, distribute pipettes of distilled water and then ammonia for them to test.

- Distribute scissors, pH paper, and color codes for the final test. Have students cut the pH paper into three pieces. Demonstrate how only one drop of liquid is placed on the end of the strip, and then the strip is immediately placed over the color code. Explain that the colors should match as closely as possible and that the pH value should be recorded as the number, not the color.

For Discussion:

- Discuss the results each group recorded. There may be some difficulty in reading the pH value at first. Have students repeat this test if necessary so they will learn how to read the code.

- Explain that the distilled water is neutral with a pH value of 7 and therefore does not change the color of the pH paper or other indicator dyes.

- Repeat the magic show and ask the students to explain what you are doing.

- Tell the students that when you did the trick using the phenol, you added vinegar, water, and then ammonia, just as they did. Explain that the phenol turned pink and remained that color after you added more water. Let them try this with their samples of phenol.

- Explain that chemists use indicator dyes like these to identify whether a substance is an acid, base, or neutral.

- Tell them that they will use some of these indicator dyes in the next lessons.

Chemical Magic *(cont.)*

Directions for Making Indicator Dyes

To the Teacher: The following directions will enable you to make the three types of indicator dyes used in these lessons. You may wish to distribute the directions for making cabbage extract indicator dye to your students since it is inexpensive, safe, and easy to make and use at home.

Cabbage Extract Indicator Dye

This indicator dye changes into a variety of colors when either base or acid substances are added to it. The color depends upon the pH value of the substance.

Materials: ¼ head red cabbage, water, pan or glass bowl, stove or microwave

Preparation:

- Chop the cabbage into small pieces.
- Put the pieces into a pan or glass bowl and cover it with water.
- Boil the water on a stove or in a microwave for about five minutes The longer the cabbage is boiled, the darker blue the water will become.
- Drain the water from the cabbage and store it in the refrigerator until ready to use.

Phenolphthalein

When a substance which is acidic is added to phenolphthalein, there is no color change. If the substance is a base, however, such as ammonia, there are bursts of bright pink color. It is an indicator of substances which are bases (alkaline).

Laxative tablets are a source of phenolphthalein since they consist mostly of this chemical. They can be dissolved in alcohol to make this indicator dye.

Materials: *uncoated laxative tablets, rubbing alcohol, flat saucer, fork, glass, storage bottle

*These may be coated but should not be chocolate. Use the least expensive laxative tablets.

Preparation:

- Place five laxative tablets on the saucer and crush them into a fine powder with the fork.
- Put the powder into a clear glass and add 1 ½ cups alcohol. Stir until dissolved.
- Let this mixture stand for about 20 minutes until the insoluble materials settle to the bottom.
- Pour the clear liquid into the bottle for storage and cap it. The sediment may be discarded.

Bromothymol Blue

If an acidic substance is added to BTB, it turns orange or yellow. If the substance is a base, the BTB will remain blue or turn back to blue if it follows an acid. It is an indicator for acids.

Bromothymol Blue (BTB) may be purchased as a powder or liquid (see Resources section). Either of these can be diluted with water. The powder is usually less expensive and goes further than the liquid. Dissolve about 1 teaspoon (5 mL) powdered BTB in a cup (¼ liter) of water. Continue adding water to the BTB solution so that it maintains its dark blue color. Test a small sample with vinegar and ammonia as you add water to avoid diluting it too much. Store this indicator in a dark closet when not in use.

Chemical Magic *(cont.)*

Testing Indicator Dyes

To the Students: You are a chemist, testing indicator dyes to find what color they turn to when an acid or base is added to them. Follow the instructions and do each test carefully.

Instructions:

1. Place the well strip with indicator dyes in them in the box below. Write "cabbage extract" on the paper in front of the three wells filled with that liquid. Next, write BTB and phenol in front of the wells containing those indicators.

Indicator Dyes

---

2. Add 2–3 drops of vinegar to the well of cabbage extract. Look at it carefully to decide what color it has become and then write the color on the record sheet.

3. Add 2–3 drops of vinegar to the BTB and record the color change.

4. Add 2–3 drops of vinegar to the phenol and record the color change.

5. Test the water and ammonia, using a new well of cabbage extract with each but the **same container** of BTB and phenol.

   You will test the pH value when the teacher distributes the pH paper at the end of the tests.

Substances	Cabbage Juice	BTB	Phenol	pH Value
Vinegar (acid)	Color:	Color:	Color:	Number:
Water (neutral)	Color:	Color:	Color:	Number:
Ammonia (base)	Color:	Color:	Color:	Number:

Is It Acid, Base, or Neutral?

Topic: Identifying known substances as acids and bases

Objective: Students test various common substances to identify them as acid, base, or neutral.

Materials:

- safety goggles
- red cabbage extract
- 8 small cups, may use 3 oz. (89 mL) bathroom cups
- pH paper and code
- 16 microscale pipettes
- 8 sets of reaction strips with 8 wells each
- transparency and 9 copies of Identification Sheet #1 (page 54)
- substances to be tested: lemon juice, baking soda solution, citrus soda pop, boric acid solution, colorless pure ammonia, clear liquid soap, vinegar, white antacid tablets (e.g., Tums®)
- 8 trays

Preparation:

- Make solutions of baking soda and boric acid by mixing as much of each substance as will dissolve in a cup of water.
- Crush several antacid tablets and dissolve them in a cup of water.
- Label eight pipettes 1–8. Make a list of the substances and then assign them a number 1–8. Fill the pipettes with the corresponding substances.
- Write the list of substances on the record sheet before making copies.
- Conduct the tests on the substances and record this data to use as an answer sheet.
- Fill the eight cups 1/2 full with cabbage extract and place a pipette in each cup.
- Place the trays, filled reaction strips, pH color code, and cups of cabbage extract on a table.

Procedure:

- Divide students into eight groups and review the lesson on using indicator dyes. Tell them that today's activity will use two of those indicators, cabbage extract and pH paper.
- Have each group send one member to the table to pick up a set of supplies for them.
- Tell students to fill each cup in the strip about 1/2 full of cabbage extract.
- Distribute the record sheet and four strips of pH paper to each group. Tell them to cut these in half so they will have eight pieces for their tests.
- Explain to students that you have filled eight pipettes with the liquids listed on their record sheets. Distribute one pipette to each group and tell them to look at the number on it.
- Tell them to test the substance by placing a few drops into a cup of cabbage extract indicator and recording its color beside the same number on their record sheet. They should test the same liquid with the pH paper and write its value on the record.
- Rotate the eight pipettes among the groups until all substances have been tested.

For Discussion:

- Compare the results by recording them on the transparency of the record sheet.
- If there are disagreements, have students retest that substance or take the answer which is in the majority.

Is It Acid, Base, or Neutral? *(cont.)*

Identification Sheet #1

Names of Group Members: _____

To the Students: You are chemists who have been given the job of identifying eight different substances to see if they are an acid, base, or neutral. Carefully test each of the substances and record the color change of the cabbage extract and the pH value. Then identify the substance by placing a √ in the acid, base, or neutral column. The chart below will help you with this job.

pH Value Range

1	2	3	4	5	6	7	8	9	10	11	12

strong acid ←—→ weak acid (neutral) weak base ←—→ strong base

Cabbage Juice Colors

orange   red   pink   purple   blue   blue green   green   yellow

strong acid ←—→ weak acid (neutral) weak base ←—→ strong base

	Substance	Cabbage Juice Color	pH Value	Acid	Base	Neutral
1						
2						
3						
4						
5						
6						
7						
8						

Mystery Substance

Topic: Identifying unknown substances as acid or base

Objective: Students will identify the same eight substances used in the previous lesson but in a different order.

Materials:

- same as in previous lessons
- Identification Sheet #1 from the previous lessons
- transparency and 9 copies of Identification Sheet #2, page 56 (one becomes the answer sheet)

Preparation:

- Change the labels on the eight pipettes to letters A–H, mixing them up so none correspond with the number they were in the last lesson. List these substances on the answer record sheet.
- Conduct the tests on the substances and record this data on the answer sheet.
- Place the supplies and new record sheet on a table.

Procedure:

- Have students get into the same eight groups as in the previous lesson.
- Explain that today they will be given the same substances they tested in the earlier activity, but they have been labeled with letters and are not in the same order as before.
- Let each group gather their supplies.
- Tell students to fill each well in the strip about ½ full of cabbage extract.
- Tell them they are to use the indicators to identify these substances as acid, base, or neutral.
- When they have completed their data, distribute their record sheets from the previous lesson to match them with the data they have just recorded on *Identification Sheet #2*.
- Tell them to write the names of the substances beside the letters on the record sheet after they have matched the data with the original set.

For Discussion:

- Ask each group to tell what the mystery substances were and write these on a transparency of the record sheet. If there are disagreements, have them discuss which answer they feel is most likely correct and why.

Follow Up:

- Write a letter to parents requesting that students bring in substances to be identified. Tell them that these will be used to help students identify acids and bases. Give some examples of substances they may want to send such as tea, cologne, or shampoo. Be sure to stress that the substances should be safe items.
- Have students test the substances they bring from home. If they bring solid substances, dissolve them in water and test the liquid. Add some substances for them to test which they did not bring from home, such as aspirin, water from a variety of sources, and saliva.

Mystery Substance *(cont.)*

Identification Sheet #2

Names of Group Members: _____

To the Students: The labels on the eight substances you identified in your previous tests have been mixed up. Run the same tests on these substances and record your results.

After conducting the tests, match this data with that on Identification Sheet #1 to see if you can name each substance.

pH Value Range											
1	2	3	4	5	6	7	8	9	10	11	12

strong acid ⟷ weak acid (neutral) weak base ⟷ strong base

Cabbage Juice Colors							
orange	red	pink	purple	blue	blue green	green	yellow

strong acid ⟷ weak acid (neutral) weak base ⟷ strong base

Mystery Substance	Cabbage Juice Color	pH Value	Acid	Base	Neutral
A					
B					
C					
D					
E					
F					
G					
H					

What Happened to Blue Lake?

Topic: Identifying pollutants in lake water

Objective: Students will identify the cause of acid pollution in a hypothetical lake.

Materials:

- safety goggles
- cabbage extract
- pH paper and code
- 8 small cups, may use 3 oz. (89 mL) bathroom cups
- distilled water
- ammonia
- vinegar
- 40 microscale pipettes
- permanent ink fine-tipped felt pen
- 4 jars with tight fitting lids capable of holding 1 cup (.25 L liquid)
- transparency and 9 copies of data sheet Pollution at Blue Lake (page 58)
- 8 trays
- 8 sets of reaction strips with four well cups each

Preparation:

- Label the jars as follows: 1991 Lake Sample, 1997 Lake Sample, Factory #1, Factory #2.
- Fill the jar marked 1991 Lake Sample with distilled water.
- Make a solution of 1 cup (.25 L) water and 1 teaspoon (5 mL) vinegar in the jar marked 1997 Lake Sample.
- Make a solution of 1 cup (.25 L) water and ¹/₂ teaspoons (2.5 mL) ammonia in the jar marked Factory #1.
- Make a solution of 1 cup (.25 L) water and 1 tablespoon (15 mL) vinegar in the jar marked Factory #2.
- Divide 32 pipettes into sets of 4 and mark them with permanent ink as follows: 1991, 1997, #1 and #2.
- Fill the cups ¹/₂ full of cabbage extract and place a pipette in each of them.
- Prepare trays with cabbage extract, pH paper and code, four water samples, and reaction strip.

Procedure:

- Divide the students into eight groups and distribute a tray of supplies to each of them.
- Use the transparency of the record sheet to discuss it with the students.
- Review the test procedure using cabbage extract and pH paper if necessary.
- Have each group fill the reaction strip wells .50 full of cabbage extract and then begin their tests.

For Discussion:

When all groups have completed their tests and summary reports, let them share their results with the class. If there are disagreements, help them resolve them, retesting if necessary.

Follow Up:

Have students do the next activity to arrive at a possible solution for saving Blue Lake.

What Happened to Blue Lake? *(cont.)*

Pollution at Blue Lake

The Problem: The fish at Blue Lake are dying. Dead fish have been found around the shore of the lake and floating on its surface. The water has a bad smell. People who have been swimming in it say it tastes sour. If something is not done soon, this may become a dead lake.

There are two factories which opened on the shore of the lake in 1992. They dump the waste water from their manufacturing process into Blue Lake. The people who live in Blue Lake City think the factories are causing the pollution.

The Mission: You are a team of chemists investigating this problem. You have been given samples of water taken from Blue Lake in 1991 and in 1997. There are also samples of the waste water from the two factories. Your team will need to test the four samples and make a report. Record the results of the tests on the chart.

---

pH Value Range

| 1 | 2 | 3 | 4 | 5 | 6 | 7 | 8 | 9 | 10 | 11 | 12 |

strong acid ⟷ weak acid (neutral) weak base ⟷ strong base

Cabbage Juice Colors

| orange | red | pink | purple | blue | blue green | green | yellow |

strong acid ⟷ weak acid (neutral) weak base ⟷ strong base

Water Specimen	Cabbage Juice Color	pH Value	Acid	Base	Neutral
1991 Lake Water					
1997 Lake Water					
Factory #1					
Factory #2					

Summary of Test Results

When the lake water was tested, we found that _____

When the waste water from the factories was tested, we found that _____

We think that the cause of the pollution is _____because

_____

Saving Blue Lake

Topic: Neutralizing a solution

Objective: Students will learn how to combine acids and bases to form a neutral solution.

Materials:

- safety goggles
- 1 qt. (1 L) cabbage extract
- pH paper and code
- 8 small cups (may use 3 oz. [89 mL] bathroom cups)
- ammonia
- vinegar
- 24 microscale pipettes

- permanent ink fine tipped felt pen
- 8 trays
- 8 1-oz. (30 mL) clear plastic cups (medicine cups)
- 8 pipettes filled with samples of 1997 Blue Lake water
- transparency and copies of Saving Blue Lake record sheet (page 60)

Preparation:

- Use permanent ink to label 8 pipettes A for ammonia and 8 pipettes V for vinegar. Make diluted solutions of 2 parts water and 1 part ammonia or vinegar. Fill the pipettes with the corresponding diluted liquids.
- Fill 8 pipettes with the 1997 Blue Lake water sample from the previous lesson.
- Fill the 3 oz. cups with the cabbage extract solution and place a pipette in each of them.
- Place the materials on trays so each group will receive a set of three pipettes (A, V, 1997 water), cup of cabbage extract, 1-oz. cup, pH paper with code, and record sheet.

Procedure:

- Discuss what students did in the "Pollution at Blue Lake" activity.
- Tell students that since acids and bases are at different ends of the pH scale, mixing them together will result in a different pH value. Explain that if just the right amounts of acid and base substances are mixed, a neutral solution can be created.
- Show a transparency of the *Saving Blue Lake* record sheet and be sure students know what is expected of them.
- Have students pour just enough cabbage extract into the 1-oz. cup to cover the bottom and then add drops of the lake water sample until a color change takes place. Have them write the color on their record sheet. Now, have them test the pH value of the solution in the cup, record it, and then circle either acid or base.
- Explain that they need to decide if they should add ammonia or vinegar. Tell them this should be added one drop at a time, recording the type of liquid added and the results.
- If students go beyond 8 trials and have not reached a pH range 6–7, have them pour out the cabbage extract solution, rinse the cup with water, and begin again.

For Discussion:

- Have the students share the results of their experiments to create a neutral solution.
- Ask what problems occurred as they worked and how these were resolved.

Saving Blue Lake *(cont.)*

To the Scientific Teams: Now that you have discovered what is causing the problem at Blue Lake, you have been hired to try to find a way to remedy the problem and make the lake water safe for fish and swimmers.

Scientific Information: A pH value of about 7 is neutral, which means the solution is balanced between an acid or base. If the solution is on the acidic side of the scale, such as pH 3, it can be changed to the neutral by gradually adding a base substance. This must be done carefully so the solution does not turn too basic. It is a bit like balancing a ruler on a pencil.

pH Value Range

1	2	3	4	5	6	7	8	9	10	11	12

strong acid ⟵⟶ weak acid (neutral) weak base ⟵⟶ strong base

Cabbage Juice Colors

orange	red	pink	purple	blue	blue green	green	yellow

strong acid ⟵⟶ weak acid (neutral) weak base ⟵⟶ strong base

The Mission: Your team needs to find out how to change the water in Blue Lake to a neutral to make it safe again.

1. Pour cabbage extract into the small cup just until it covers the bottom. Add drops of 1997 Blue Lake water until it changes color. Record the color on the chart. Test the cabbage extract solution with the pH paper and record it. Draw a circle around acid or base.

2. Add a drop of ammonia (base) or vinegar (acid) to bring the water toward neutral. After each addition, look at the color of the cabbage extract and see if it is changing toward neutral. Add ammonia or vinegar, depending upon what the solution needs to bring it to neutral.

3. Check the pH value after adding 3 drops and, if it is between 6 and 7, it is neutral and you are finished. If it is higher or lower, continue to add drops until it is at a neutral level.

Neutralizing Blue Lake						
1997 Blue Lake Sample: Cabbage Juice Color:_____ pH_____				acid		base
#drops acid	base	Color	pH	acid	neutral	base

Identifying Acids and Bases

To the Teacher: This is a performance-based assessment in which students do four tasks, applying the skills and knowledge they have gleaned from the activities in this unit.

Task	Activity	Materials
1	Identifying acids and bases	• 2 pipettes labeled #1 and #2 filled with diluted solutions of ammonia and vinegar (used in "Saving Blue Lake") • reaction strips with four cups • small cup of cabbage juice with pipette
2	Finding a pH value	• 2 pipettes labeled #3 and #4 filled with diluted solutions of ammonia and vinegar • pH paper and code
3	Identifying a solution	• results of test from Identification Sheet #1 • 3 pipettes labeled A, B, C, each filled with substances tested in this activity representing an acid, base, and neutral • small cup of cabbage juice with pipette • reaction strips with four cups
4	Problem solving	• paper and pencil

Administering the Assessment:

- Make copies of the correct results of Identification Sheet #1 (page 54) to be used in task #3.
- Gather the materials needed for tasks 1–3 and create at least two centers of each task.
- Place the materials for the tasks on tables and/or desks. Label the center 1, 2, or 3. If needed, place cardboard screens on the desk or tabletop around each center for privacy.
- Distribute the answer sheet to the students and discuss the tasks. Tell them that they may be starting at center #1, 2, or 3 but that they will eventually rotate to each center.
- Explain that they must clean up their centers after they are done so everything is just as they found it. They need to rinse reaction wells with water and replace the paper towel, if needed.
- Establish a rotation system that will enable students to spend no more than 10 minutes at a center. This may be spread out over a day or more to accommodate all students.
- An optional method would be to assess the entire class at the same time if sufficient materials are available to make more than one center for each task. All students can rotate to these centers on schedule within a 30-minute period if multiple centers are arranged.
- Write task #4 on the board for students to work on it when not at one of the centers.

Task #4: You have eaten a food which is acidic, and your stomach feels upset. What would help make your stomach feel better, something which is an acid or a base? Explain your answer.

- If the entire class takes the assessment at one time, students can complete task #4 while waiting for others to finish until the 10-minute work period ends.

Grading Rubric:

- Conduct the tests for the three tasks and use the results for correcting the students' work.

Identifying Acids and Bases *(cont.)*

Name: _____   **Date:** _____

pH Value Range											
1	2	3	4	5	6	7	8	9	10	11	12
strong acid	←	→	weak acid		(neutral)		weak base	←	→	strong base	
Cabbage Juice Colors											
orange	red		pink	purple	blue	blue green		green		yellow	
strong acid	←	→	weak acid		(neutral)	weak base		←	→	strong base	

Task 1

Test the liquids in pipettes # 1 and #2 to find if they are acid or base.

1. Use the pipette and fill two cups in the reaction strip half full of cabbage extract.
2. Place drops of liquid #1 into a cup of cabbage extract until it changes color and record the results.
3. Put drops of liquid #2 into the other cup of cabbage extract until it changes color and record the results.

Liquid	pH Value	Acid	Base	Neutral
1				
2				

Task 2

Test the two liquids with the pH paper to see if they are an acid or a base and then record the results.

1. Place one drop of liquid #1 on the tip of the pH paper and match the color on the pH color code.

Liquid	pH Value	Acid	Base	Neutral
1				
2				

2. Put a drop of liquid #2 on the tip of another pH paper and match the color on the pH color code.

Task 3

The liquids in pipettes A, B, and C were used in the activity in which you identified eight substances as acid, neutral, or base. Test liquids A, B, and C and record. Match these with the identified substances on Identification Sheet #1 to find out what these mystery substances really are.

Mystery Substance	Cabbage Juice Color	pH Value	Acid	Base	Neutral
A					
B					
C					

Flying a 737

This simulation section was written by a commercial pilot, Bruce Young, who began to prepare himself for that career while in the third grade. He started by building model airplanes and reading everything he could find on the topic of flight and airplanes. Bruce spent countless hours practicing on flight simulation software on a home computer. He talked to pilots and went to air shows and frequently visited a small local airport. At age 17 he began to take flight lessons at that airport; by 18 he had a private pilot's license. After high school, he attended an aeronautical university to continue the training he needed. Following a year of more training, he was assigned as a first officer (co-pilot) on a commuter airline in Miami. He became a captain (pilot) with that airline at age 23, the minimum age limit for this level. His studies were not over at this point, however, since every time he is assigned to a new aircraft, he spends a week or more at "ground school." This includes reading and learning everything about the new airplane. This includes practice sessions on a simulator designed to teach how to fly the new plane. Finally, there is a rigorous flight test in the aircraft with a specialist who evaluates his flight skills.

The number of flight hours in command of an aircraft are carefully recorded by pilots. This is an official record of their flying experience, and it is carefully maintained. Hundreds of flight hours are required before moving into positions of flying more sophisticated aircraft. The minimum of 1,500 flight-in-command hours is required to fly a 737.

Flying commercial planes is a great responsibility. Pilots continue to train for different aircraft and are retested throughout their careers. They must also be physically fit. A physical examination is required by the Federal Aviation Administration (FAA) semi-annually. If pilots have failing eyesight, diabetes, heart trouble, or other health problems which may affect their flying performance, they could lose their license or have it suspended until the problem is eliminated.

The airline industry is rapidly expanding, and the mandatory age for pilots to retire is 60. Many of today's pilots are close to retirement age. This leaves openings for younger people to become pilots. More women are now taking up careers as pilots.

The activities within this simulation section are designed to show students how fascinating and exciting flying can be. They will also find that pilots need to know about the parts of an airplane and how they fly. Pilots must be able to read complicated navigational maps, use a new alphabet, and speak a special language. They have to understand weather conditions so they can know what to do when flying in bad weather.

The pilot does not fly a large plane alone. Other crew members include the first officer, navigator, flight attendants, and ground crew. The plane needs guidance from air traffic controllers on the ground. Students perform a play about flying a commercial 737 jet from Miami to Atlanta. The script includes the official language used by the flight crews and ground controllers throughout the world. It enables students to briefly become members of a flight or ground crew involved from takeoff to landing this 737.

How Do Planes Fly?

Topic: Aerodynamics

Objective: Students will identify the parts of a plane which help it to fly.

Materials:

- transparencies of pages 65–68 *Parts of an Airplane* and *How an Airplane Turns*; *Basic Cockpit Controls and Instruments*; *Pitch, Roll, and Yaw*; *Airplane Wings* and *Airplane Tails*
- lined paper for each student
- Optional: airplane pictures and books

Preparation:

- Display the airplane pictures and books before this lesson so students will have the opportunity to look at them and become excited about learning more about flying.

Procedure:

- Ask the students to gather in small groups and share what they know about airplanes.
- Create a list of the things students know about airplanes.
- Distribute lined paper and have the students write three things they want to know about airplanes.
- Have students tell some of the things they want to learn about airplanes.
- Show the transparencies to provide more information about airplanes. (You may want to make copies of these pages for the students to enclose in a journal.)
- If available, show pictures of actual airplanes that are examples of the wings and tails shown in the drawings.

Follow Up:

- Have students add the new information they learned about planes to their papers.
- Let students make an airplane journal which will contain their own work, as well as copies of information and worksheets from this study.
- Do the next activity in which students make an airplane from paper.

How Do Planes Fly? *(cont.)*

Parts of an Airplane

This is a drawing of a light, propeller-driven airplane. The basic parts are the wing, body, tail, landing gear and engine. The ribs and spars in the wings add structure without much weight. Ailerons, flaps, rudder, and stabilator are movable parts used to control the plane. All planes have most of these same parts. Jets do not need a propeller since their engine gives the thrust needed to push the plane through the air.

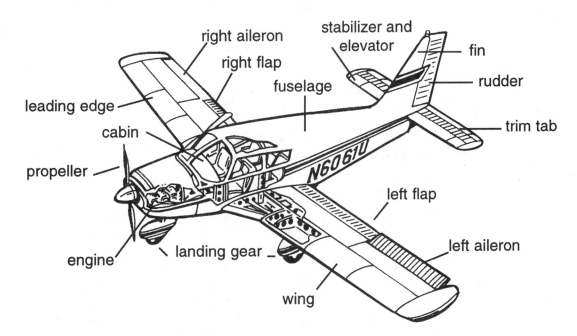

How an Airplane Turns

The pilot uses several controls to turn the airplane. This plane is beginning a right turn so the pilot raises the right aileron to bank (tip) the plane to the right. The rudder is also turned to the right to keep the plane's nose steady. Lift on the left wing is increasing so the plane is pulled around the turn. The plane loses lift as it turns so the pilot raises the elevator and increases the engine power.

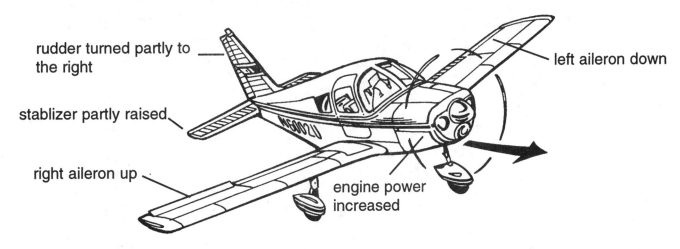

How Do Planes Fly? *(cont.)*

Basic Cockpit Controls and Instruments

Both the wheels used by the pilot and copilot (first officer) control the ailerons and elevator. Their pedals operate the rudder. The throttle controls the engine power and speed. Flight instruments such as the air-speed indicator and altimeter are used to keep the plane on course. Engine instruments such as the oil-pressure gauge and tachometer measure engine operations.

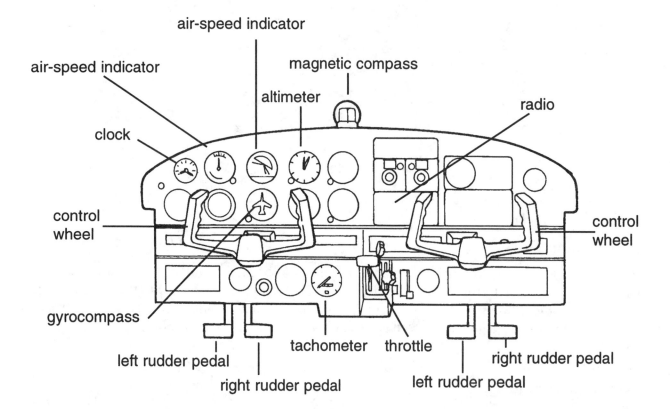

How Do Planes Fly? *(cont.)*

Pitch, Roll, and Yaw

An airplane has three basic movements, (1) pitch, (2) roll, and (3) yaw. To make a plane **pitch**, the pilot lowers the elevator by pushing the control wheel forward or raises it by pulling the wheel back.

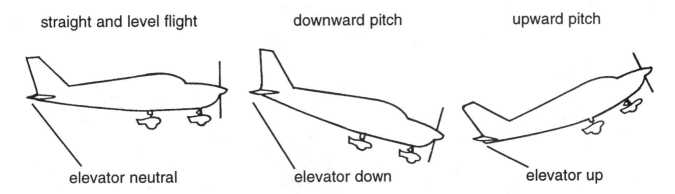

straight and level flight   downward pitch   upward pitch

elevator neutral   elevator down   elevator up

The pilot uses the ailerons to make the plane **roll** or **bank**. To make a left bank, the pilot turns the control wheel left which raises the left aileron and lowers the right one.

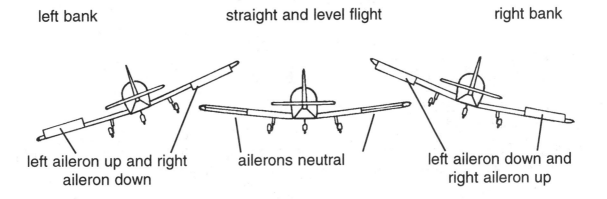

left bank   straight and level flight   right bank

left aileron up and right aileron down   ailerons neutral   left aileron down and right aileron up

To make a plane **yaw**, the pilot operates the two pedals that control the rudder. Pressing the right pedal swings the rudder to the right, and the left pedal swings it to the left.

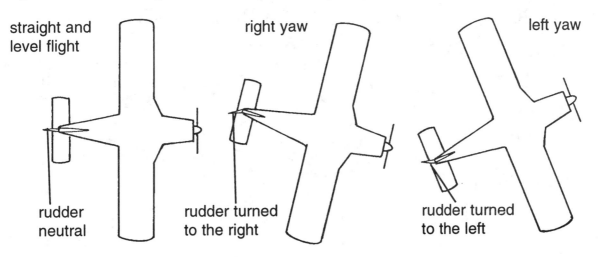

straight and level flight   right yaw   left yaw

rudder neutral   rudder turned to the right   rudder turned to the left

How Do Planes Fly? *(cont.)*

Airplane Wings

The shape of an airplane wing depends upon the type of plane. Straight wings are best for both high and low speeds. Swept-back or swept-forward wings are used on high-speed jets. Delta wings are used on jets also. They provide high speed and great lift.

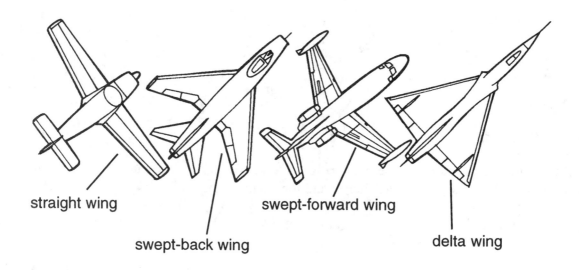

straight wing

swept-forward wing

swept-back wing

delta wing

Airplanes Tails

The vertical tail (fin and rudder) may be at a right angle to the fuselage or swept back. Some planes have two (twin) or three fins. The drawings below show a variety of airplane tails.

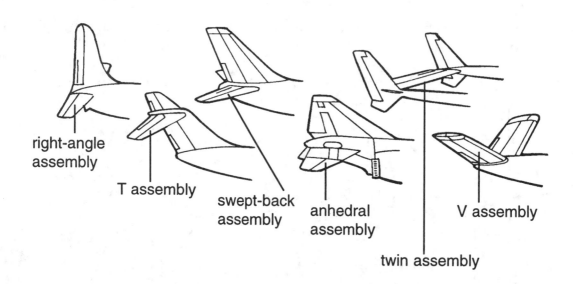

right-angle
assembly

T assembly

swept-back
assembly

anhedral
assembly

V assembly

twin assembly

Making a Paper Plane

Topic: Aerodynamics

Objective: Students will construct simple paper airplanes and test-fly them to learn how they perform.

Materials:

- worksheet, How to Make a Paper Plane (page 70)
- 8.5" x 11" (22 cm x 28 cm) stiff paper for each student
- clear tape

Preparation:

Optional: Gather ideas for making a variety of paper planes. These may be found in book stores and also in *Science Is . . . A Sourcebook of Fascinating Facts, Projects, and Activities* (see Resources section).

Procedure:

- Let students tell some of the things they have already learned about airplanes.
- Explain that before there were planes with engines, there were gliders. Tell them that they are going to make a glider from paper and then test-fly it.
- Distribute the worksheet and paper to the students so they can make their own paper airplanes.
- Test-fly the paper airplanes inside the room first, to avoid wind. Let students modify their planes to help them fly better and then retest them.
- Take the planes outside and flight-test them again.

For Discussion:

Have students discuss what they learned about the design of paper airplanes. Which design could fly the longest, highest, straightest?

Follow Up:

- Have students design a variety of paper airplanes.
- Do the lesson on constructing an airplane from balsa wood.
- Let students write a brief research paper about airplanes and early flight.

CE National, Inc.
Resource Library

Making a Paper Plane *(cont.)*

How to Make a Paper Plane

There are a wide variety of ways to make a simple airplane from paper. You are building a glider. Gliders are very light planes that have no engines. They use their long wings and air currents to help them fly.

Follow the instructions below to construct a simple paper plane.

1. Fold the paper lengthwise down the middle. Unfold it and lay it flat again.

2. Fold one of the corners over to the center fold, and then fold the other corner the same way.

3. Fold the corners a second time so they meet in the center fold.

4. Fold the two sides together with the corners inside. Make the wings by folding the top portion of each side down to the center fold.

5. Place a small piece of clear tape across the center when the wings come together. Tape the end of the paper under the wings together.

6. Launch this glider by holding it at the back, below the wings. Think of it as a dart and try to throw it as straight as possible. Test-fly the glider inside a room before trying it outside.

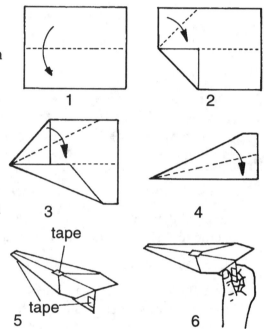

After test-flying your paper plane, try to make it better. Some suggestions are listed below. Try each of these changes one at a time to see what happens when you test-fly the plane again.

- Snip off or bend about .5" (1 cm) of the nose. Add a paper clip to the plane's nose to add weight and then try more than one paper clip.

- Cut flaps (ailerons) in the wings. What happens if only one flap is bent up and one down?

- Bend up the wing tips to act as vertical stabilizers. What happens when these are bent down?

- Add rudders by cutting flaps in the vertical stabilizers and bending them in different directions.

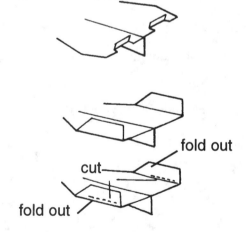

What Keeps an Airplane Up?

Topic: Bernoulli's Principle

Objective: Students will discover how air flows over an airplane's wings can create lift

Materials:

- transparency of What Keeps an Airplane Up? (page 72)
- 2" x 4" (5 cm x 10 cm) strip of paper for each student

Procedure:

- Let students tell what they have learned about how a plane flies.
- Distribute the strips of paper to the students.
- Have them gently curl them around their fingers, just to create a slight curve to them.
- Tell them to hold one end of the paper near their lips and then gently and steadily blow across the top of the paper. Have them report what happens (the paper begins to rise as air passes over it and then drops again when they stop blowing).
- Tell them that airplanes use this same principle to help them stay in the air.
- Show them the transparency and use it to explain how planes can stay up in the air.

For Discussion:

- **Lift:** Have students try their experiment with the strips of paper again and explain how this is used by airplanes. Let them explain what this has to do with how an airplane stays in the air.
- **Drag:** Have students wave a piece of paper to feel air pressure against it. Explain how the paper is being slowed by the drag of air pressure against it.
- **Gravity:** Drop something to the floor and have students explain why it fell rather than floated. (Gravity from Earth pulls everything towards its center.) Ask them to tell how this affects airplanes.
- **Thrust:** Ask them to explain where their paper airplanes got their thrust. (The muscles in their arm as they launched the plane were like the force of an engine in a plane.)

Follow Up:

Construct a balsa wood airplane.

What Keeps an Airplane Up? *(cont.)*

Forces on an Airplane

Four forces act on a plane during its flight.

- **Lift** is upward force created by air pressure.

- **Thrust** is forward push provided by the propeller or jet engine.

- **Gravity** is a downward pull caused by Earth's gravity.

- **Drag** is air resistance caused by friction by air rubbing against the plane.

Airplanes are designed so that lift and thrust are stronger than the pull of gravity and the drag of the air. The faster the plane moves, the greater it can overcome gravity. Airplanes are shaped like an arrow so they can cut through the air easily and reduce the drag from the air. The front edges of the wings are curved to give lift.

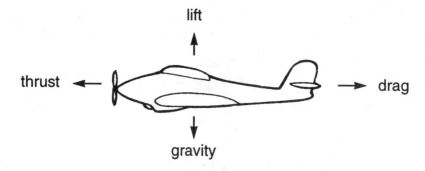

What Keeps an Airplane Up?

A Swiss mathematician and physicist, Daniel Bernoulli (1700–1782) explained how fluid (liquid or gas) behaves when it is moving in a horizontal direction. He discovered that if the speed of the gas or liquid decreases, pressure also decreases. This explains how airplane wings create an upward force called *lift*. Airplane wings are shaped with a curved top surface and flat bottom surface. As the engines force the curved wings through the air, air passing over the top has further to go than the air below which is traveling in a straight path. The air over the top of the wing must go faster than the air below the wing so all air reaches the edge as the same time. The pressure below is therefore greater than the air pressure above because of the different air speeds, and the plane is lifted.

The wings of an airplane can be extended or retracted, using adjustable flaps. When extended, the flaps increase the curvature of the wings on the upper side and provide greater lift for takeoff and landing.

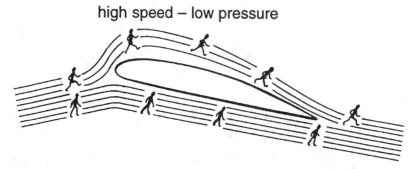

Building a Working Model Airplane

Topic: Applying aerodynamics

Objective: Students will construct a balsa wood airplane and test-fly it.

Materials:

- *balsa wood airplane for each pair of students (or per student if funds permit)
- masking tape
- file cards
- scissors

- transparencies of Parts of an Airplane (page 65), and Pitch, Roll, and Yaw (page 67) from "How Do Planes Fly?"

*available at most hobby or toy shops, be sure to get the rubber-band-powered model

Preparation:

Assemble one of the airplanes and test-fly it.

Procedure: (This lesson may be spread over more than one day.)

- Review the parts of an airplane and how a plane turns, using the transparency.
- Tell students that they are going to construct their own model airplanes and test fly them.
- Divide the students into pairs and distribute the balsa wood airplanes.
- Have them carefully lay out each piece and punch out the airplane parts from the sheet of balsa wood. Be sure to tell them not to begin assembling their planes.
- Tell the students to build their airplanes along with you so they can discuss the parts as they add them to the model.

Instructions and Discussion:

- The first component is the *fuselage*, the main structure of the aircraft. This carries the passengers and cargo and in some aircraft the fuel and landing gear. This is also where all flight surfaces attach to the aircraft.
- Place the wing on the fuselage as shown in the model instructions. The *wing* is the main lifting surface of the aircraft. Some planes also use the fuselage to provide some of the lift, but the wing is the primary source. Most wings also hold the fuel, landing flaps, and ailerons.
- Demonstrate flying your plane (but not the students') with just the wing attached. Notice how it just flips over. Now, place the *stabilizer* on your plane. This helps to stabilize the plane in the pitch axis and prevents it from tumbling end over end. It is also the attachment point for the elevators that help pitch the nose of the plane up and down in flight.
- Your aircraft is now stable in pitch, but it will be difficult to fly in a straight line. Just as an arrow or a throwing dart needs stabilizing feathers on its tail, an aircraft needs a vertical surface to stabilize it and keep it flying through the air in a straight line. This is the vertical stabilizer, and it should be placed just above the horizontal stabilizer. The vertical stabilizer also houses the rudder, which is the movable control surface that helps steer the nose of the airplane left and right.
- Have the students add these parts to their models. Show the transparency of the Pitch, Roll, and Yaw and review the parts of the wing needed to help the airplane maneuver.

Building a Working Model Airplane *(cont.)*

Instructions and Discussion: *(cont.)*

- An airplane needs some form of propulsion. If your airplane is a glider (i.e., paper plane), then the propulsion is provided by your arm when you throw the plane. However, that propulsion source is removed as soon as the plane leaves your hand. If you have a rubber-band-powered propeller on your plane, attach it now. Point out that the blades of the propeller act just like wings and are also curved. As they spin through the air, they create lift in a horizontal direction that serves to pull the aircraft along through the air.

- Before you go outside to test fly your aircraft, have students write their names on them. They may also want to decorate them with colored pens. Now, they are ready for the big test.

- Demonstrate with your plane how to wind the propeller and release the plane. Explain that, just like an airport, air traffic control determines where and how many planes may approach the field. Tell them you are the air traffic controller and will tell them when to launch their planes. Have them work in groups of four or five and take turns lining up to allow space for their planes to fly.

Follow Up:

- After every student gets a turn, have all students try variations with their airplanes. Have them collect data on how their airplanes perform.
  - √ Launch the plane at an upward angle.
  - √ Launch them from the ground, using pavement as a runway.
  - √ Find out how many turns of the propeller seems to give the best performance.
  - √ Design your own test.

For Discussion:

- Return to the classroom and discuss the results of the test flights.

- Discuss various ways to modify the aircraft so they can improve performance.

- Use the transparencies which show the elevators, rudder, and ailerons on the plane and how these are used to turn the plane, as well as pitch, roll, and yaw.

- Use small pieces of file cards approximately 1" x .5" (2.5 cm x 1 cm) and tape.
  - √ **Elevators**—Tape the elevators to the trailing edge of the horizontal stabilizer, and bend them up or down, depending on how you want the nose of the plane to move.
  - √ **Rudder**—Tape one piece of paper to the vertical stabilizer and bend it slightly left or right to move the aircraft now in flight.
  - √ **Ailerons**—Tape the pieces of card to the trailing edges of the wings and move them to control the plane in flight. They must move in opposite directions to roll the aircraft.

- Rather than all groups making all three modifications, have each select one of the modifications.

- After modifications are made, return to the launch site and repeat the test flights. Be sure to take the tape, cards, and scissors to the field so repairs and changes can be made.

Flight Vocabulary

Topic: Communication between ground and flight crews

Objective: Students will learn that the vocabulary and alphabet used between ground and flight crew are different from any other language.

Materials:

transparency and copy of Pilot Alphabet worksheet (page 76) for each student

Background Information:

When flight first began, there were so few planes that the pilots made their own decisions about take-offs and landings, as well as where in the sky to fly the plane. As more planes took to the air, traffic regulations had to be established to avoid running into each other, just like cars on the highway. This required communication between the controllers on the ground and the pilot and/or first officer. The first radios used in airplanes were poorly made, and the conversations were often broken up by static or distance so some words would be missing. For this reason, a special vocabulary was developed so the communication could be clearly understood and kept as short as possible.

Many planes were, and still are, identified by numbers and letters. On the old radios, letters may not have come through clearly, thus a special alphabet was developed for communicating. Each letter was given a word. If the pilot was reporting a plane's identification code of XYZ123, he/she would say, "X-ray, Yankee, Zulu, 1, 2, 3." By using the word in place of the letter, there was more of a chance that the message would come through clearly.

Despite the fact that today's radios are much better, the vocabulary and alphabet used to communicate while flying remain the same. Messages are kept very brief since there are so many pilots who need to speak to the ground controllers at the same time. The English language is used for communication with flight crews and controllers by all airports in the world.

Procedure:

- Share the background information with the students.

- Distribute the worksheet and show the transparency copy of it. Have the students complete the first one with you so they will understand how to complete the rest. For example, the pilot says, "This is Alpha Romeo 1 7 6," Which means, "This is plane AR176."

- Let students complete the worksheet and then have them check each other's code to see if it is correctly written.

Follow Up:

Conduct the simulation of the commercial airliner flight.

Flight Vocabulary *(cont.)*

Pilot Alphabet

To the Student: The alphabet shown below is used by ground controllers and flight crews. If they are saying just letters, have them say the word instead of the letter. For example, DE533 would become: Delta, Echo, 533, and taxiway J would be "taxiway Juliet."

A	Alpha	J	Juliet	S	Sierra
B	Bravo	K	Kilo	T	Tango
C	Charlie	L	Lima	U	Uniform
D	Delta	M	Mike	V	Victor
E	Echo	N	November	W	Whiskey
F	Foxtrot	O	Oscar	X	X-ray
G	Golf	P	Papa	Y	Yankee
H	Hotel	Q	Quebec	Z	Zulu
I	India	R	Romeo		

Now it is your turn to talk using *pilot's language*. Change the following to show how a pilot would say them or what they mean.

Pilot's Alphabet	
Pilot says or hears . . .	**Which means . . .**
• This is Alpha Romeo 1 7 6.	•
•	• This is plane ZC345.
• take taxiway papa.	•
•	• take taxiway H.

Invent an identification code for your own imaginary plane. Write it below as you would say it and then as it would be said by a pilot:

My plane's identification code is ____ ____ ____ ____ ____

As a pilot I would say, _____

Flying a 737

Topic: Simulating the flight of a 737 commercial airliner

Objective: Students will learn how an airline flight takes place by acting out a simulated airline flight.

Materials:

- transparency copies of the two route charts from Miami to Atlanta (pages 88–89)
- transparency of the Glossary of Flight Terms (page 80)
- poster-sized copy of 737 cockpit panels (pages 90–92) mounted on display board
- transparency of Simulated Flight Cast Members (page 79)
- copies of the script (pages 81–87), one for each cast member and the teacher
- cart on wheels with prepackaged snacks for passengers
- overhead projector and screen (optional: poster-size copies of airport diagrams and route charts)
- optional props: 6 headsets with wire to simulate microphones for Pilot, FO, OPS, TWR, DEP/ARR, and ENR, telephone, strips of paper or cloth for seatbelts

Preparation:

- Arrange the classroom with three tables in the front of the room. The middle table will be for the Captain (CAPT) and First Officer (FO). Place a chair behind the CAPT for Flight Attendant (FA) #1; the FA #2 sits behind the passengers. The FAs should only have chairs and face the passengers. Passengers in a 737 are seated in two rows, three seats across. Those students who are not part of the cast will arrange their chairs in this configuration. Seatbelts may be simulated by paper strips taped to the chairs of the crew and passengers. The crew also have shoulder belts for additional safety. Headsets for the CAPT and FO, a microphone for the FA #1, and a telephone for FA #2 may be added as props.
- The Announcer (ANN) and Airline Operations (OPS) will sit at another table. At the third table should be the Tower (TWR), Departure/Arrival (DEP/ARR), and Enroute Controller (ENR) positions. The cockpit panel poster is placed upright on the table in front of the Captain and First Officer. The room arrangement is shown below.

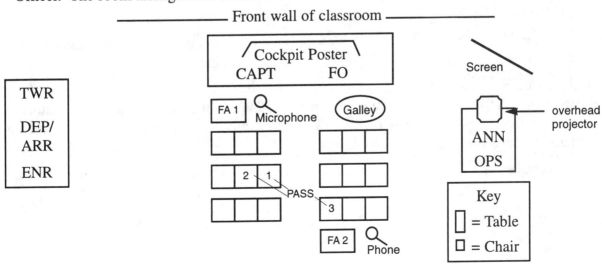

Flying a 737 *(cont.)*

Preparation: *(cont.)*

- Make poster-sized copies of the three 737 cockpit panels. Match the edges and mount the panels on display board. Place these panels of instruments upright in front of the flight crew.

- Make transparency copies of the two route maps. The announcer will show the position of the plane by placing a silhouette of the plane (shown below) in the location indicated in the script. The route is shown as a bold line (airway) extending from Miami to Atlanta.

- Makes copies of the script for each cast member. Highlight one role on each copy to make it easier for the cast members to follow the script.

- Prepare snacks by placing them in a baggie or other package so they are easy for the flight attendants to serve to the passengers during the simulation.

For Discussion:

- Ask students if they have ever flown on a commercial aircraft. Have them share these experiences in small groups and then have some tell the whole class.

- Ask the students how many would like to be part of a flight crew someday.

- Show the transparency of the cast members and explain what each one does. (Do not explain the roles of the passengers so that this will be a surprise as the simulation proceeds.)

- Show the transparency of the Glossary of Flight Terms to the students to prepare them for the script of the simulation.

Procedure:

- Choose students to play the roles. Distribute a copy of the script to each cast member.

- Allow time for the crew members to read their parts of the script to become familiar with it. Be sure they do not tell any details of the script to non-cast members to keep it as a surprise.

- Items embedded within the script between < > are directions for the cast members to follow.

- Have the cast take the appropriate seats within the setting you arranged in the classroom.

- Have all non-cast members seated in the passenger seats, facing forward.

- Set the scene by telling students this flight begins with the aircraft at the departure gate, with the cabin door just being closed prior to pushing back from the gate.

- Tell the students that this is a simulation of the events that can unfold on a typical airline flight from Miami, Florida, to Atlanta, Georgia. This simulation will encompass all of the factors involved on an everyday flight and will show all of the different people involved during the flight, including flight crew and ground controllers.

- Conduct the simulation by letting cast members read their scripts aloud and act out the parts.

Flying a 737 *(cont.)*

Simulated Flight Cast Members

(CAPT)   Captain: responsible for the entire aircraft, the crew, and the passengers

(FO)   First Officer: assists the Captain on the flight deck

(FA 1) and **(FA 2)**   Flight Attendants: responsible for the safety of the passengers and the security of the cabin, also assist and serve passengers to make their flight more comfortable

(TWR)   Tower Air Traffic Controller: in the Control Tower on the airport; oversees all local air traffic within 5 miles of the airport

(DEP/ARR)   Departure/ Arrival: air traffic controller in a radar facility; oversees all air traffic departing from and arriving at the airport

(ENR)   Enroute Air Traffic Controller: in a radar facility; oversees all air traffic flying enroute from one airport to another airport

(OPS)   Operations Airline personnel: responsible for dispatching flights and helping the flight crew from the ground during the entire flight

(ANN)   Announcer: acts as a narrator during the simulation

(PASS 1)   Passengers on the flight

(PASS 2)   Passengers on the flight

(PASS 3)   Passengers on the flight

Flying a 737 *(cont.)*

Glossary of Flight Terms

Runway 9L via taxiway Papa—Runway 9 left on taxiway P

40 past—40 minutes past the hour

turn left heading 060—turn left 60°

V1 Rotate V2—V1 is take off commitment speed; rotate is takeoff speed; V2 is safe climb speed in the event of an engine failure

flight level 250—flight at altitude 25,000 feet

VOR—navigational locations scattered around the world, except on the ocean; transmit a very high frequency within a certain area to help pilots locate their position

jet airway—an imaginary highway for planes to follow in the sky

TransAir 531 is with you out of 500 feet for 5000 feet—After take off, the pilot uses the new frequency to let the ground controller know the altitude change so the plane can be picked up on the radar screen.

weather in Atlanta 500 overcast—clouds 500 feet above the ground; means poor visibility above 500 feet from the ground

Flight Management Computer—performs many navigation functions during flight; tells autopilot and crew where to fly the plane

frequency 134.2—airport transmission frequency

ATC—Air Traffic Control

ILS (Instrument Landing System)—electrical signals on the runway to guide pilots in, especially necessary when visibility is poor in bad weather

flaps 15—flaps set at 15°, flaps extended from the wings to adjust the plane's speed; flaps set at 15°; will slow the plane and altitude

12 o'clock—imaginary clock face (parallel to the ground); 12 o'clock position, straight ahead; 3 o'clock 90° to the right; 6, directly behind; 9, left 90°

thrust reversers—exhaust from the engines force, quickly slows the plane during landing

80 knots—1 knot = 1.15 miles; 80 knots per hour, 92 mph

Note: Read each number separately. For example, 40 is read 4–0 and 500 is read 5–0–0.

Flying a 737 (cont.)

Flight Simulation Script

To the Cast: Items shown between < > are directions for the cast members to follow. Read each number individually. For example, 531 is read as 5–3–1, not five hundred thirty-one.

FA 1	Captain, the cabin is secure.
CAPT	Thank you. *<to the First Officer>* Let's run the Before Engine Start Checklist, please.
FO	Roger, Captain. Cabin Doors...
CAPT	Secure.
FO	Cockpit Checks . . .
CAPT	Complete.
FO	Auxiliary Power Unit . . .
CAPT	Running.
FO	Engine Start Switches . . .
CAPT	Select engine two.
FO	Engine two is turning.
CAPT	Let's call for pushback and taxi.
FO	Miami Tower, this is TransAir 531, ready for pushback and taxi from Gate Foxtrot 11.
TWR	TransAir 531, Miami tower. You are cleared for pushback and taxi to Runway 9L via taxiway Papa.
FO	Roger, taxiway Papa to 9L, TransAir 531.
CAPT	Okay, let's turn engine one.
ANN	As the cockpit crew runs through their engine start procedures and pre-taxi checks, the cabin crew is responsible for briefing the passengers on information they would need to know in the unlikely event of an inflight emergency.
FA 1	Ladies and gentlemen, welcome aboard TransAir flight 531 with non-stop service to Atlanta, Georgia, continuing on to Chicago. Our flying time to Atlanta today will be one hour and 50 minutes. I would like to take this opportunity to briefly explain to you some of the safety features of our Boeing 737 aircraft. Please direct your attention to the Flight Attendants standing in the aisle and follow along with the safety briefing cards in the seatback pocket in front of you.
FO	Miami Operations, 531.
OPS	Go ahead, 531.
FO	Yes sir. We're off the gate at 40 past, 157 passengers, on time.
OPS	Roger, 531; have a good flight.
TWR	TransAir 531, your takeoff sequence is behind the 737 ahead of you. You're number two for departure on runway 9L.
FO	Roger, thank you.
CAPT	*<Talking on the Public Address (PA) system to the passengers in the back>* Welcome aboard from the flight deck. My name is Capt. (your last name), and assisting me today is First Officer (first officer's last name). We're number two for departure here at the end of the runway, so we should be airborne within a couple of minutes. I'd like to ask the Flight Attendants to please be seated for departure.

Flying a 737 *(cont.)*

Flight Simulation Script *(cont.)*

ANN	The flight is now cleared into position on the runway and is ready for takeoff.
TWR	TransAir 531, runway 9L, after takeoff turn left heading 060, and you are cleared for takeoff.
FO	Roger, left 060, cleared for takeoff, TransAir 531.
CAPT	Set max takeoff power.
FO	Takeoff power set, engine instruments all look good . . . V1 . . . Rotate . . . V2
CAPT	Positive rate of climb *<pause>* landing gear up.
FO	Gear's coming up, come left to heading 060.
CAPT	Roger, left to 060. Flaps up.
TWR	TransAir 531, contact departure, and have a good flight.
FO	Roger, over to departure, see you later *<pause>* Miami departure, good morning. TransAir 531 is with you out of 500 feet for 5000 feet.
DEP/ARR	Good morning TransAir 531. Climb and maintain flight level 250, and turn left heading 360.
FO	Roger, climb to 250, and left to 360.
CAPT	Set climb power; enable the autopilot.
ANN	The flight is now climbing up toward its cruising altitude of 35,000 feet. As they climb, the air begins to get smoother, and the Captain turns off the seat belt sign. The Flight Attendants then begin their in-flight snack and drink service. They are currently passing over the La Belle VOR. *<place the plane silhouette on route map #1 transparency over this spot and leave it on the screen for a few minutes>*
FA 1	Ladies and Gentlemen, the Captain has turned off the fasten seat belt sign. Feel free to get up and move about the cabin. However, if you are going to remain in your seat, we ask that you do keep your seat belts fastened. We will be coming around shortly to provide you with our snack service.
	<Flight attendants begin to serve the snacks to passengers.>
DEP/ARR	TransAir 531, contact Miami center now on frequency 134.2.
FO	Roger, 134.2. Have a good day *<pause>* Miami center, TransAir 531 is with you climbing out of flight level 180 for flight level 250, heading 360.
ENR	Good morning 531. Turn left to heading 290 to intercept Jet Airway 73, and climb to flight level 350.
FO	Roger, left to join J73, climbing to flight level 350.
CAPT	Miami OPS, 531.
OPS	Go ahead, 531.
CAPT	Yes sir . . . 531 was out at 40 past and off at 47. We'll be on time into Atlanta.
OPS	Roger 531. The current weather in Atlanta is 500 overcast, 3 miles visibility, with a chance of thunderstorms developing in the area in the next hour.
CAPT	Okay, we'll keep an eye on it with your help. Thanks a lot.*<to the first officer>* Looks like the weather is going to get worse by the time we get to Atlanta. Let's keep an eye on it with OPS, and think about going to our alternate, Birmingham, if it gets too bad.
FO	Okay.

Flying a 737 (cont.)

Flight Simulation Script (cont.)

ANN	Flight 531 has leveled off at its cruising altitude of 35,000 feet. The flight is on autopilot at this point, and the Flight Management Computer will fly the aircraft along its entire route. The crew will monitor the progress of the flight through their displays in the cockpit, and the computer will alert them if there is any malfunction in any system. There will be occasional communications with Miami Center as they fly across Florida. These communications will contain information about nearby traffic, changes in weather conditions, and radio frequency changes as they move from one Air Traffic Controller's airspace to the next. The flight is now passing the Lakeland VOR. *<move the plane on transparency #1 and leave it on the screen>*
ENR	TransAir 531, you're leaving my airspace *<pause>* contact Atlanta Center on 125.7. Have a good day.
FO	Roger, Atlanta on 125.7. See you all later. *<pause>* Atlanta Center, good morning. TransAir 531 is with you at flight level 350.
ENR	TransAir 531, Atlanta Center. Good morning.
FA 1	*<on the intercom from the cabin to the cockpit>* Captain, it's getting too bumpy back here for us to do our snack service. Is there any way you can give us a smoother ride?
CAPT	Sure thing. *<to first officer>* Let's ask Atlanta Center if there is a smoother ride at another altitude.
FO	Atlanta Center, TransAir 531. We're getting some turbulence up here. Is there any smoother altitude up ahead?
ENR	Yes, sir. Flight level 290 is reported to be a lot smoother ride, and so is flight level 370. It's your choice.
CAPT	The headwind would be too strong at 290; let's climb to 370.
FO	Atlanta, 531 would like to climb to flight level 370.
ENR	Roger 531, you're cleared up to flight level 370.
CAPT	*<on the PA to the passengers>* Ladies and gentlemen, from the flight deck *<pause>* we're going to climb up to 37,000 feet to get out of this turbulence. *<pause>* Air traffic control has told us it should be a lot smoother at that altitude.
OPS	Flight 531, operations.
CAPT	Go ahead OPS, this is 531.
OPS	Yes sir; it looks like Atlanta weather is continuing to go down; they're currently reporting one and a half miles visibility and a ceiling of 500 feet. Will your fuel status allow you to go to Birmingham in case you can't get into Atlanta?
CAPT	Yes, we have more than enough fuel. We'll continue on to Atlanta and try to get in there. If the weather goes below the minimums, we'll go to Birmingham. But we really want to try to get into Atlanta.
OPS	I understand. I'll keep you updated on how the weather is doing.
CAPT	Thank you. *<to first officer>* OPS says the weather in Atlanta is still going down. It's still 500 feet and one and a half miles. I told them we'd still try to get in before the weather closes. Do you feel comfortable with that?

Flying a 737 *(cont.)*

Flight Simulation Script *(cont.)*

FO   Sure. The weather is still above our approach minimums. There's really no reason why we can't still try to get in.

ANN   Flight 531 is approaching the Atlanta airport, and they begin their descent around 120 miles out. All during the descent, the flight crew is keeping a close eye on the weather conditions in Atlanta. If the weather goes below a 200-foot ceiling and a half-mile visibility, they will have to go to Birmingham, Alabama, and wait for the weather to come back up so they can return to Atlanta as soon as possible. As the flight nears its destination, the cabin crew is busy cleaning up the cabin and securing everything for arrival. They are currently over the Valdosta VOR. *<move the plane to transparency #2 to this spot and leave it on the screen>*

ENR   TransAir 531, contact Atlanta Approach Control on 119.75.

FO   Roger, Approach on 119.75. *<pause>* Atlanta Approach, TransAir 531 is with you descending from flight level 240 to 15,000 feet.

DEP/ARR   Good morning, TransAir 531, Atlanta Approach. Continue descent to 8,000, and fly heading 040.

FO   Roger, heading 040, down to 8,000; TransAir 531.

CAPT   Atlanta operations, 531.

OPS   Go ahead, 531, this is Atlanta.

CAPT   We'll be on the ground in 15 minutes.

OPS   Roger 531, 15 minutes. You'll be at gate 47. Be advised the weather here is at minimums, 200-foot overcast and one-half mile visibility, with a thunderstorm approaching the airport in about 20 minutes.

CAPT   Roger, thank you. *<to first officer>* Looks like we're right at minimums. and they have a thunderstorm approaching right about the time we'll be landing. Let's get ready to go to Birmingham.

FO   Roger, Captain.

PASS 1   *<holding chest in pain and coughing>*

PASS 2   *<waving frantically to flight attendant>* This passenger next to me is having trouble breathing and says his chest hurts.

FA 2   *<talking over the microphone>* Is there a doctor on board?

PASS 3   I'm a doctor.

FA 2   *<speaking to the doctor>* Please come with me to check a passenger who is having trouble breathing.

PASS 3   *<speaking to sick passenger and taking his/her pulse>* How bad does your chest hurt?

PASS 1   It feels like someone is sitting on my chest, and it hurts to breathe.

PASS 3   *<speaking to FA 2>* His pulse is racing. I think he should go to a hospital as soon as possible. I don't think this is a heart attack, but we need to have it checked.

FA 1   *<on intercom to the captain>* Captain, we have a problem back here. We have an elderly passenger who just started complaining about chest pains. We found a doctor on board, and he's helping the passenger. He said we have to get him on the ground as soon as possible and get him to a hospital.

CAPT   I understand. Is his condition serious?

Flying a 737 *(cont.)*

Flight Simulation Script *(cont.)*

FA   I don't think so. He just says his chest hurts a lot, and the doctor suggests we get him to a hospital so they can run some tests on him.

CAPT   Okay, we'll try to get down as soon as possible. We should be on the ground in about 10 minutes. *<to the first officer>* Looks like we've got a sick passenger in the back, and we'll need to get down in a hurry. I think we ought to try to get into Atlanta first, even with the weather being right at minimums. Ask the controller for priority handling to get us to the airport.

FO   Do you want me to declare an emergency?

CAPT   Negative. Let's just get to the airport without any delays and try to get there before that thunderstorm.

FO   Roger. The radar shows the storm about 10 miles north of the airport. I'll call ATC *<on the radio>* Atlanta approach, TransAir 531 is requesting priority handling to the airport.

DEP/ARR   TransAir 531, do you wish to declare an emergency?

FO   Negative, sir. We have a sick passenger who needs to get on the ground immediately. We just need a short cut to the approach.

DEP/ARR   Roger, 531. Turn right heading 090, and descend to 5,000. Maintain your best forward speed to the outer marker, and you're cleared for the ILS approach to runway 12. Atlanta is currently reporting one-half mile, indefinite ceiling with rain showers on the field.

FO   Roger, right 090, down to 5,000, and best speed to the marker; 531 is cleared for the ILS approach runway 12. Thank you very much.

CAPT   Atlanta OPS, 531.

OPS   Go for Atlanta OPS.

CAPT   Yes, sir; 531 has a passenger on board complaining of chest pains. We'll be on the ground in 7 minutes, and we'll need an ambulance to meet us at the gate.

OPS   Roger 531. Be advised the thunderstorm will most likely be at the field by the time you arrive.

CAPT   I understand, but we need to get this guy on the ground soon. Our radar shows that we might make it just before the storm gets there, so we're going to try to get in now.

OPS   Roger, 531. Call us when you're on the ground.

FO   Captain, we've been cleared for the approach, and we're number one for the airport.

CAPT   Thank you. I'm going to leave the autopilot on for the approach since the weather is so bad. Let's run through the final approach checklist.

FO   Roger that. *<on the PA to the cabin>* This is the first officer. We are preparing to land in Atlanta. We will be experiencing some bumps since it is turbulent weather there at this time. We do not want to take time to go to an alternate airport due to the illness of one of the passengers. It will be safe to land in Atlanta, but all passengers should follow the instructions of the flight attendants. *<pause>* Flight Attendants, please prepare the cabin for arrival.

Flying a 737 *(cont.)*

Flight Simulation Script *(cont.)*

ANN   As Flight 531 begins its final approach, the Flight Attendants are making their final preparations for landing. They are checking to make sure all loose articles in the cabin are secure and that all passengers have their seatbelts fastened. They are also attending to their sick passenger. In the meantime, Operations in Atlanta is busy calling an ambulance to meet the flight, and Atlanta Approach Control is busy moving other aircraft out of the way so that flight 531 can be the first aircraft to the airport. With the weather going below minimums, Approach Control has to place other aircraft into a holding pattern until the weather can get better or work on sending them to their alternate airports.

DEP/ARR TransAir 531, contact Atlanta Tower on frequency 118.4. Good day.

FO   118.4, have a good day. *<pause>* TransAir 531 *<pause>* Atlanta Tower, TransAir 531 is with you for runway 12.

TWR   Roger TransAir 531. You're cleared to land on runway 12. The runway visibility is one-half mile, and there are rain showers on the northeast corner of the field.

FO   Roger, 531 is cleared to land.

CAPT   There's the outer marker. Landing gear down, flaps 15.

FO   Gear's coming down, flaps are selected 15 and indicating.

CAPT   1000 feet above the field.

FO   Roger. Landing gear is down and locked, three green lights. Flaps are selected 15 and indicating. You're on speed, and your sink is 800 feet per minute. Everything looks good.

CAPT   Flaps landing, please.

FO   Flaps landing are selected *<pause>* and indicating. Landing checks are complete.

CAPT   I can see the ground now but not the runway. It looks like the rain shower is just approaching the runway now.

FO   It sure is raining pretty hard. *<pause>* Do you think we'll be able to see the runway?

CAPT   Well, if we don't see it by 200 feet, we'll go around and hold until the weather clears a little.

FO   500 feet. On speed, sink 700.

CAPT   Runway is in sight, 12 o'clock.

FO   Roger, I see it.

CAPT   I'm taking over for a manual landing *<pause>* autopilot is disconnected.

FO   200 feet, on speed.

CAPT   Touchdown. Deploying thrust reversers.

FO   Two green lights . . . reversers are deployed *<pause>* 80 knots, come out of reverse now.

TWR   TransAir 531, nice job. When you're clear of runway 12, taxi to your gate.

FO   Roger, TransAir 531 will taxi to gate 47.

CAPT   After landing checklist, please. Looks like we just got in before the worst of the storm. Thank goodness!

FO   It sure does. You did a great landing, Captain, especially for all that weather. After landing checks in progress.

CAPT   Thanks. There's our gate, and it looks like the ambulance is already there.

FO   Atlanta OPS, 531 is on the ground, going to gate 47. It looks like the ambulance is there waiting for us . . . thanks a lot.

Flying a 737 *(cont.)*

Flight Simulation Script *(cont.)*

OPS   Roger, 531. Nice job getting in here before the weather.

<All passengers clap and cheer for the great landing.>

FA   *<on intercom>* Captain, our passenger is doing much better. He said the pain in his chest is gone, but the doctor still suggests he go to the hospital.

CAPT   Thank you very much. There is an ambulance waiting for us at the gate.

FA   Thank you, Captain.

ANN   As flight 531 taxis into the gate, the flight crew prepares to shut down the engines and secure the aircraft. They will then help prepare the plane for it's next flight up to Chicago, which is scheduled to leave within an hour. The next day, they will fly this same trip in reverse order, ending up back in Miami. Hopefully, the weather will be better tomorrow and no emergencies occur.

Flying a 737 *(cont.)*

Route Map Part 1

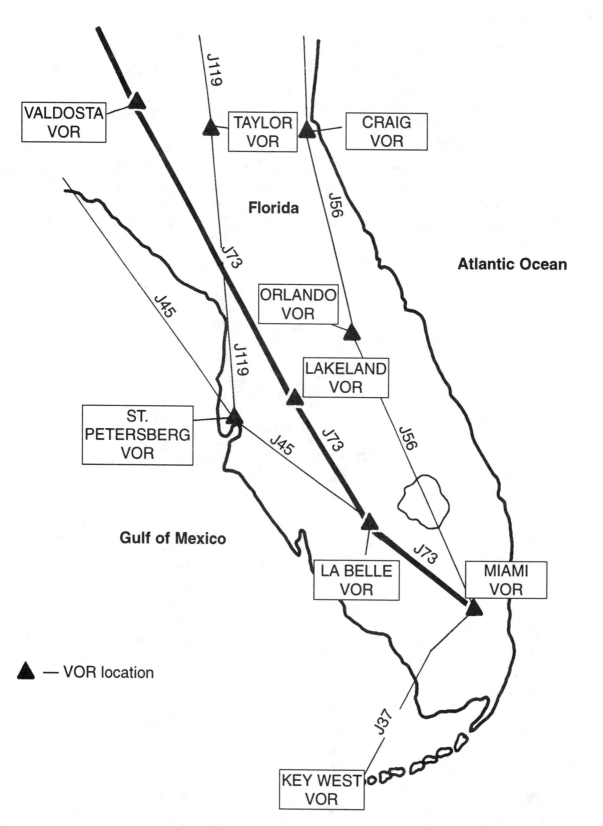

Flying a 737 *(cont.)*

Route Map Part 2

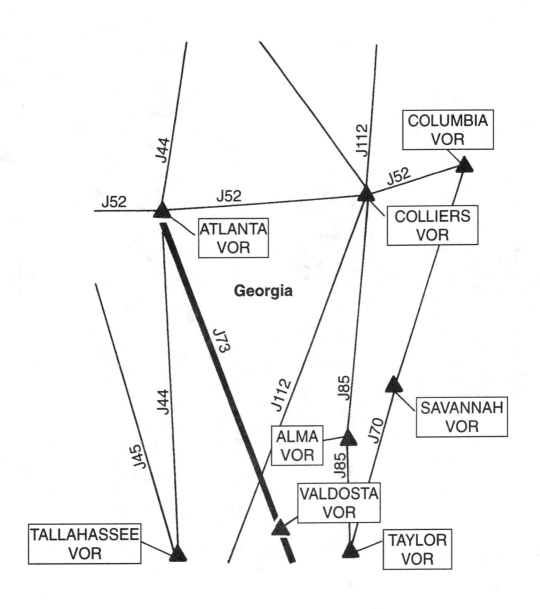

▲ — VOR location

Flying a 737 *(cont.)*

B–737 Cockpit

Captain's (left) Side Panel

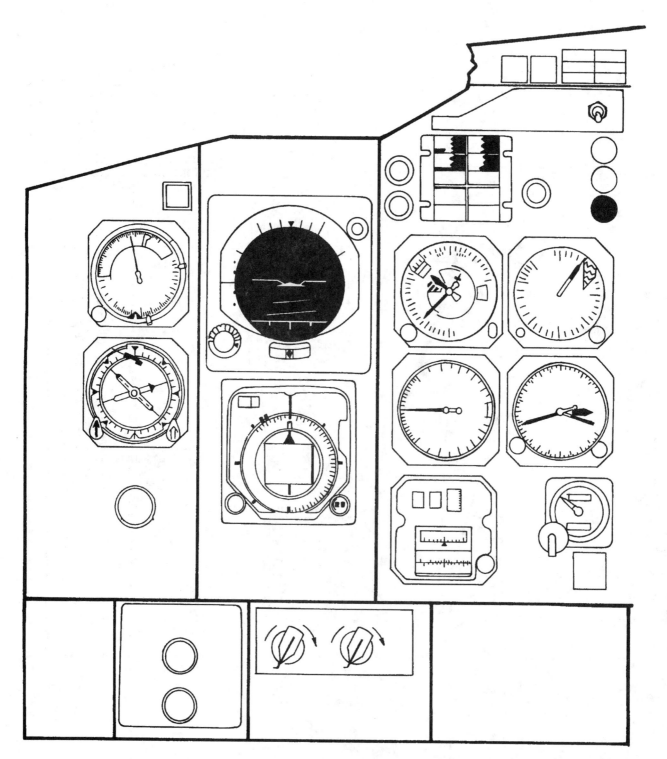

Flying a 737 *(cont.)*

B–737 Cockpit

Center Panel

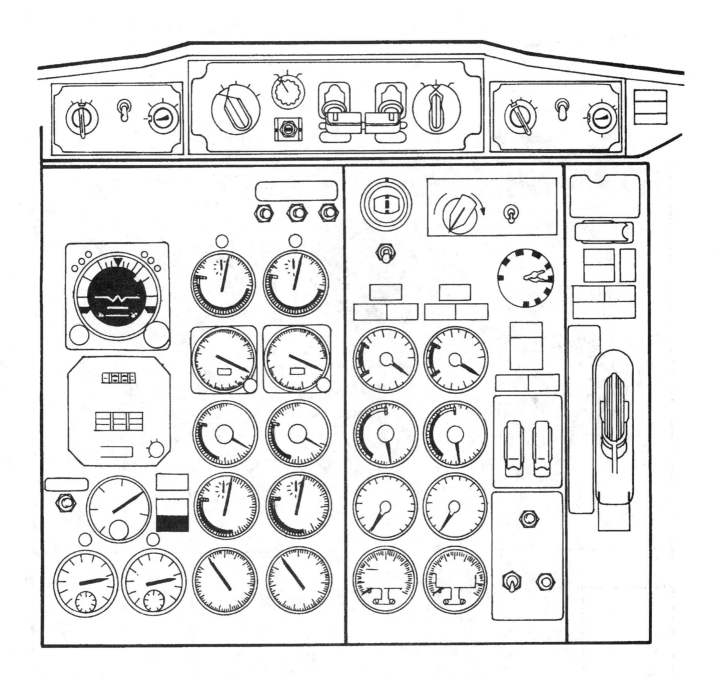

Flying a 737 (cont.)

B–737 Cockpit

First Officer's (right) Panel

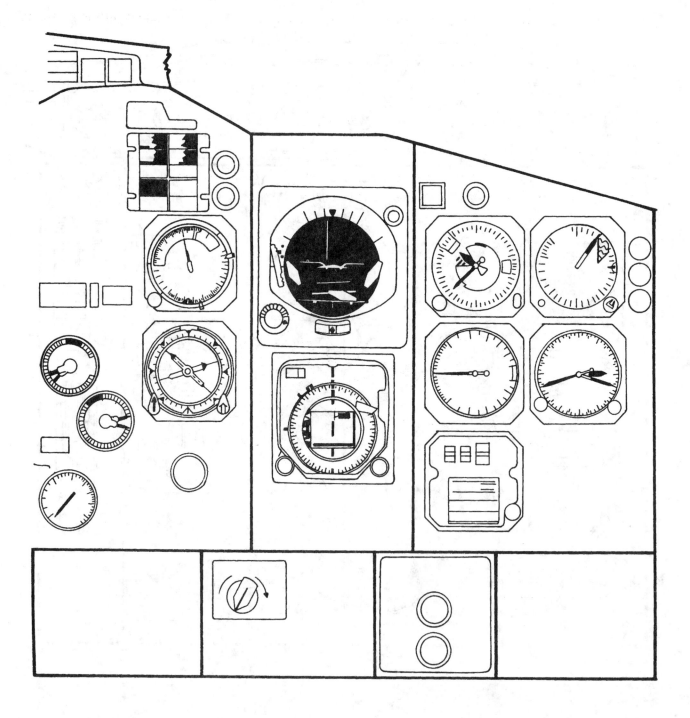

Working for an Airline

Name: _____ **Date:** _____

To the Student: Circle one of the airline positions listed below which you think might interest you as a future career. Tell something about what you would do in this job. Explain why you feel you would be the right person for this position.

pilot   first officer   flight attendant   air traffic controller   operations airline personnel

_____

_____

_____

_____

Label the airplane to show where these parts are located: fuselage (body), wings, propeller, flaps on the wings (ailerons), vertical and horizontal stabilizers (tail sections), elevators and rudder (on the stabilizers).

Explain how a plane is able to fly. Include drawings on the back to illustrate what you write.

_____

_____

_____

Resources

Books and Periodicals

Bosak, Susan V. *Science Is . . . A Sourcebook of Fascinating Facts, Projects, and Activities.* Scholastic Canada LTD, 1991. (May be ordered through NSTA–see below.) This teacher's guide is packed with great easy-to-do activities, including several on building paper planes.

Cobb, Vicki. *Chemically Alive: Experiments You Can Do at Home.* J. B. Lippincott, NY, 1985. Easy-to-do chemistry experiments, using materials commonly found at home or in a pharmacy.

Chemical Tests. Science and Technology for Children, Carolina Biological Supply Co., 2700 York Road, Burlington, NY 27215 (800) 334-5551. Teacher's guide introducing basic principles of chemistry to middle elementary grades.

Great Explorations in Math and Science (GEMS) series by Lawrence Hall of Science, University of CA, Berkeley, CA 94720 (also available from Carolina Biological Supply Co.). This series of teacher's guides includes: *Acid Rain, Chemical Reactions,of Cabbages and Chemistry.*

VanCleave, Janice. *Chemistry for Every Kid.* John Wiley & Sons, 605 Third Ave. NY, NY, 10158. (also available from Carolina Biological Supply Co. and NSTA). This book is part of a science series by the popular author and includes 101 simple-to-do chemistry experiments.

Related Materials

Carolina Biological Supply Co., 2700 York Road, Burlington, NY 27215 (800) 334-5551. Suppliers of biological equipment, specimens, books, and other materials.

Delta Education, P.O. Box 3000, Nashua, NY 03061-3000 (800) 442-5444. Supplies science materials such as bromothymol blue and pH paper, safety goggles, magnets, etc.

Electric Circuits Teacher's Guide by Science and Technology for Children. Available from Carolina Biological, which also supplies wire, bulbs, battery holders, and bulb sockets.

Flinn Scientific Inc., P.O. Box 219. Batavia, IL 60510 (800) 452-1261. Supplies bromothymol blue, pH paper, test tubes, microscale pipettes and well strips, safety goggles, as well as other chemicals and science equipment. Call for a free chemical catalog reference manual.

Insect Lore, P.O. Box 1535, Shafter, CA 93263, (800-LIVE BUG). Supplies live specimens including ants, butterfly larvae, and silkworm eggs. Also offers variety of books and science materials. Call for free catalog.

National Science Teachers Association (NSTA), 1840 Wilson Blvd., Arlington, VA 22201, (800) 722-NSTA. Publishes annual catalogs of science education suppliers, and books available through NSTA.

Scientific, Edmund Scientific Co., 101 East Gloucester Pike, Barrington, NY 08007-1380 (800) 728-6999. Supplies magnets and a wide variety of other science supplies.

Answer Key

Will the Bulb Light? (page 29)

1. (N) The bulb needs to make contact with the negative (–) end of the battery or have another wire attached to the side of the metal base of the bulb and the (–) end of the battery.

2. (Y) The wire contacts the side of the metal base of the bulb, and the bottom of the bulb base is contacting the battery. The other end of the wire contacts the (–) end of the battery.

3. (N) The wire cannot touch both ends of the battery. This causes a short circuit so electricity flows through the battery into the wire and back into the battery rather than through the light bulb.

4. (Y) This will work when the wire contacts the metal base of the bulb at its tip and the metal base contacts the battery. It will not work if both the wire and the base of the bulb contact the battery. Notice that the battery is reversed from those in 1–3 but makes no difference.

5. (Y) This is possible to do if the metal bases of all bulbs are in contact with each other, the bases of the bulbs make contact with the battery, and the wire touches one of the metal bases on the side. The current then transfers from the wire, passes through each bulb and into the battery, and goes through the wire at the bottom to complete the circuit.

6. (N) The wire must contact the (–) end of the battery.

7. (N) The bulb must contact the (–) end of the battery.

8. Accept any answer which is different from those shown in drawings 1–7 and makes the required contacts to light the bulb.

Constructing Circuits (page 33)

1. When bulb 1 is unscrewed, the other lights remain illuminated.

2. When any of the bulbs is unscrewed, the others remain illuminated.

3. The electricity flows from the battery, to the socket, through the bulb, and back to the battery. Electricity flows in this same way for each of the sockets.

4. When bulb A is unscrewed, the other two lights go out also.

5. When any of the bulbs are unscrewed, all other bulbs go out.

6. The electricity flows from the battery to the first socket, through the bulb, and directly on to the next sockets and bulbs, returns to the battery, and continues the flow.

7. When one light goes out in a parallel circuit, the others remain illuminated. When one light goes out in a series circuit, none of the other lights are illuminated.

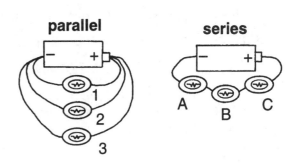

8. The electricity in a parallel circuit will bypass the bulb which is not illuminated and still light up the other bulbs. This cannot happen in a series circuit since the link is broken when any one of the bulbs burns out and electricity cannot flow to the next one and back to the battery.

Answer Key *(cont.)*

Will the Circuit Work? (pages 35 and 36)

1–4. Circuit A does not work; circuits B, C, and D will work. Electricity looks for the shortest path, which in circuit A is between the two batteries. This is called a short circuit, and electricity never flows through the bulbs. To correct this, one battery can be removed or turned around. When the batteries are linked in this way, the bulb will light but not burn any more brightly with two batteries. Electricity flows from battery to battery and then to the light bulb without giving the full power of two batteries.

5. Circuits B and C are parallel.

6. Circuit D is a series.

Mystery Connections Record (page 41)

The lighter lines are the actual wires which should appear on the record of the connections students make. The dark lines in the diagrams show possible connecting lines students will show on their record sheets but which are not actual wires. All of the lines for each box show the way electricity can flow to make the connections between the brads.

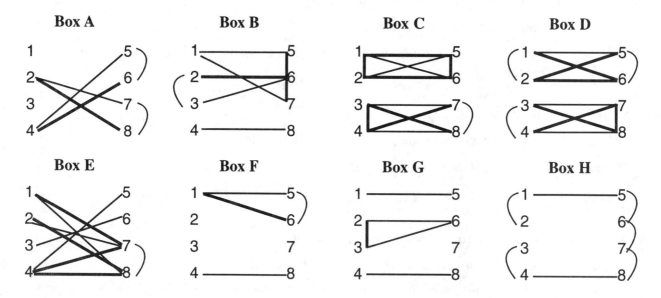

Pilot Alphabet (page 76)

- This is Alpha Romeo 1 7 6 = This is plane AR176

- This is Zulu Charlie 3 4 5 = This is plane ZC345

- Take taxiway papa = Take taxiway P

- Take taxiway Hotel = Take taxiway H